Seeing Beneath the Soil

Prospecting methods in archaeology

Anthony Clark

To Richard Atkinson

First published 1990 by
B. T. Batsford Ltd

Revised paperback edition 1996

Reprinted 2000, 2001
by Routledge
11 New Fetter Lane, London EC4P 4EE

Simultaneously published in the USA and Canada
by Routledge
29 West 35th Street, New York, NY 10001

Routledge is an imprint of the Taylor & Francis Group

Transferred to Digital Printing 2003

© 1990 Anthony Clark

Printed and bound in Great Britain by
Biddles Short Run Books, King's Lynn

British Library Cataloguing in Publication Data
A catalogue record for this book is available from the British Library

Library of Congress Cataloging in Publication Data
ISBN 0–415–21440–8

Contents

	Preface	7
	Preface to the second edition	8
	Acknowledgments	9
	Acknowledgments to the second edition	9
1	The development of archaeological prospecting	11
2	Resistivity	27
3	Magnetometry	64
4	Magnetic susceptibility	99
5	Other methods	118
6	Choice of method: choice of site	124
7	Interpretation and presentation	132
8	Survey logistics	158
	References	165
	Glossary	168
	Supplement	171
	Index	188

Preface

When I began this book, I thought it would be fairly easy to produce a general worldwide review of geophysical prospecting in archaeology; but as my allowance of pages and illustrations was consumed, I realized how big the subject had grown, and how impossible it was to encompass fully in one reasonably sized book.

Therefore the book has become largely a manifesto for the methods we have adopted and developed in Britain. Perhaps this is no bad thing, because I know of no other country where archaeological prospecting is so routinely and widely applied. Clients have come to expect extensive and economical surveys and this has led, by a process of quite natural selection, to the development and survival of the most practical techniques, and the experience is worth sharing. This is not to say that we owe no debt to colleagues abroad, as I hope I shall make clear.

I suppose all authors of this type of book hope to strike just the balance that will inform the professional and appeal to the general interested reader. I have certainly borne in mind the needs of the smaller-scale and amateur practitioner as well as the professional, and have gone to war on mystification and jargon – which I don't understand anyway. Of course, technicalities have to be included when dealing with such a subject, but I do not think that much of the sense of the story will be lost by glossing over them, at least on a first reading. Whatever the technicalities, most of us ultimately admit that the appeal of archaeological prospecting, which can be uncomfortable and tedious, is the romance and enduring excitement of this peculiarly civilized detective work – restoring to visibility the settlements of long-forgotten people who also knew our lands, by means of technology beyond their dreams.

I suspect that a groan will go up from many of my colleagues when they see some of the illustrations, which have already been well exposed in the archaeological literature. I can only excuse myself by saying that some of the old examples simply do remain the best and clearest, and it is perhaps useful to gather them together in a definitive way. Others have long lurked in my mind as potential examples, and it is good also finally to make use of these.

Writing about his book *A Brief History of Time*, Stephen Hawking said that he was told by his publisher that every equation he included would halve his readership. Even so, he felt that he had to run the risk of including $E = mc^2$, but that its familiarity might make it scarcely detectable. I hope that the same will apply to $I = V/R$. Other equations do creep in, but perhaps mentioning metal detectors and dowsing will help to redress the balance.

My debt to others for kindness and help over the years is enormous; I am really only reporting on their achievements. Many are acknowledged in Chapter 1, and I will reserve this paragraph for special thanks to my colleagues and successors in the Geophysics (now Archaeometry) section of the Ancient Monuments Laboratory, notably Alister Bartlett (still a colleague in the world outside), Andrew David and David Haddon-Reece, who have brought great intelligence and ingenuity to the subject and have endured dreadful hardships at my behest. They have been extraordinarily helpful and generous in providing illustrations from their own material.

This book has proved impossible to compartmentalize neatly. It was essential to have chapters on survey logistics and the interpretation and presentation of results, but material that would have fitted into them had to be introduced in earlier chapters to illustrate and explain the survey techniques. Magnetic susceptibility as a survey method seemed to follow naturally after magnetometer survey, yet had to be introduced in the earlier chapter. Phosphate surveying seems to pop up all over the place. I am grateful to my brother, Perry Clark, who guided me through these difficulties, explaining that they arose from the fact that the different components of the subject were related among themselves as well as to the central, essential themes of survey and presentation, a structural problem that was not unique.

The book includes some results of original research, especially in resistivity, which have not been previously published.

Guildford, June 1989

Preface to the second edition

The need for a second English edition of this book has arisen fairly quickly. It therefore seemed appropriate to leave the original layout largely untouched apart from corrections, and to put most new material into a supplement at the end. This revised version also forms the basis of the first Japanese edition.

The first publication of the book in 1990 coincided with the appearance in Britain of the enormously influential government planning policy guidance note, PPG16, which advises regional planning authorities to require developers to fund archaeological investigation as a precondition of the granting of planning permission. This has changed the emphasis of much archaeological prospecting from a mainly site-specific role to a search for archaeological sites on a landscape scale. Road and rail corridors many kilometres in length, and areas for mineral extraction, often 50 hectares or more in area, must be searched economically and quickly for the presence of remains. This need, and steady improvement of the pictures that can be produced by archaeological geophysics, have led to an increasing demand for instrumental prospecting as part of the standard routine of archaeology, and there are now a number of small companies in Britain and abroad specialising in providing this service. The role of the Ancient Monuments Laboratory has shifted from that of main practitioner increasingly towards development, setting standards and monitoring the work of others.

Dealing with large evaluation exercises has mainly fallen to magnetometry and magnetic susceptibility, because of their speed and strength of response to former human activity. The use of magnetic susceptibility has especially developed as a result, and Chapter 4 has had to be considerably complemented.

The book has become adopted as a text for college courses. Therefore I have included a little more mathematical substance on resistivity in the supplement.

Guildford, January 1996

Acknowledgments

Material for this book has been gathered from many sources. Quotations have been acknowledged in the text, and I am grateful to the following for permission to use illustrative material, much of it copyright.

Academic Press Ltd: Figs. 34, 39, 54, 55, 78; Martin Aitken: Figs. 7, 79; Edward Arnold Ltd: Fig. 49; Richard Atkinson: Fig. 2; Bartington Instruments Ltd: Fig. 81(A); Alister Bartlett: Figs. 105, 107(B, C); B. T. Batsford Ltd: Fig. 87; Martin Bott: Fig. 49; British Gas PLC, Fig. 98; Cambridge Committee for Aerial Photography: Figs. 4, 107(A); Clarendon Press, Fig. 79; Fondazione Lerici: Figs. 14, 25(A), 37, 104; Geonics Ltd: Fig. 25(B); Geoscan Research: Fig. 53; P. Goodhugh: Fig. 43; David Graham: Fig. 108; Edward Hall: Fig. 51; Colin Heathcote: Fig. 63; Historic Buildings and Monuments Commission (English Heritage), Ancient Monuments Laboratory and Publications Section: Figs. 55, 57, 58, 72, 75, 77, 78, 85, 86, 88, 94, 96–104, 106, 107(B), 110; Tsuneo Imai: Fig. 92; John Mead: Fig. 3; Megger Instruments Ltd: Fig. 1; National Monuments Record, Royal Commission on the Historical Monuments of England: Figs. 42, 84; Mark Noel: Fig. 48; Northampton Development Corporation: Fig. 54(A); Jeffery Orchard: Figs. 54(C), 65, 66; OYO Corporation: Fig. 92; Frank Philpot: Figs. 13, 14; Plessey Company: Fig. 13; Research Laboratory for Archaeology, Oxford: Figs. 9, 51, 64; Julian Richards: Fig. 86; Graham Ritchie: Figs. 73, 74; Royal Commission on the Ancient and Historical Monuments of Scotland: Figs. 73, 74; Irwin Scollar: Figs. 8, 9; Sheffield University: Fig. 48; Society of Exploration Geophysicists: Fig. 92; Surrey University: Fig. 20; Vivien Swan: Figs. 14, 15; Thames & Hudson: Fig. 21; Thamesdown Archaeological Unit: Fig. 105; Michael Tite: Fig. 64; Trust for Wessex Archaeology: Fig. 86; David R. Wilson: Fig. 87.

Acknowledgments for the second edition

I must add my gratitude to the following for material they have allowed me to use in the second edition of this book.

Buckinghamshire County Museum and Department of Transport: Fig. 124; Cotswold Archaeological Trust and Wates Built Homes Ltd: Fig. 125; Historic Buildings and Monuments Commission (English Heritage), Ancient Monuments Laboratory and Publications Section, and Neil Linford: Fig. 120; Yasushi Tanaka and STG Corporation, Shanghai: Fig. 127; Stratascan, RPS Clouston and Severn-Trent Water plc: Fig. 123.

I am also grateful to J.C.E. Jennings for helpful suggestions.

Chapter 1
The development of archaeological prospecting

Describing his work on Handley Down, Dorset, in 1893-5, Lieutenant-General Augustus Pitt Rivers wrote:

> The ground to the west of Wor Barrow was examined to ascertain if any trace of habitations could be found, but nothing of any kind could be seen upon the surface. The pick was then used to hammer on the surface, and by this means, the Angle Ditch was discovered. The sound produced by hammering on an excavated part is much deeper than on an undisturbed surface, a circumstance worth knowing when exploring a grass-grown downland, though not applicable to cultivated ground.

Thus the great pioneer of scientific archaeology first described the use of a 'geophysical' method of archaeological detection, in Volume IV of his *Excavations in Cranborne Chase*. Elsewhere in this volume, he says that the 'flat side' of the pick was used, presumably the head with the handle vertical. This simple technique, which came to be called 'bosing', still has a place in archaeology (Atkinson, 1953) although in practice it can be quite difficult to use. It is similar in principle to modern sonic reflection techniques, which have still to be fully developed for archaeological prospecting.

Another relevant early technique, geochemical rather than directly instrumental, was phosphate detection. Especially associated with Scandinavia and developed in the 1920s by the Swedish agronomist Arrhenius (1929), its potential seems to have been first appreciated in Egypt in 1911 by F. Hughes (Russell, 1957). It is still being used and refined. Responding to enhanced phosphate concentrations in soil, derived from the waste products of man and animals, it can define general areas of habitation and in some cases can even provide quite detailed evidence of site usage. It will be discussed later in this book.

*

In the valley of the middle Thames, under the brooding twin hill of Wittenham Clumps, lie ancient villages and towns whose roots lie in some of the first settlements of the English. Beneath the soil are hidden in profusion vestiges of these and much older human endeavours, including the pits and ditches of mysterious sacred enclosures built at least four millennia ago. Levelled by the plough, these only become visible to modern eyes from the air in certain seasons, through their effect on the growth of crops.

One particular field at Dorchester (now a worked-out gravel pit turned into a boating lake) was the scene of famous events. At one end of it lay the Big Rings, a Neolithic monument of a type known as a 'henge', a ritual enclosure related to Stonehenge. The first site not on chalk to be discovered from the air, it was confirmed by O. G. S. Crawford and R. G. Collingwood by excavation in 1927. At the other end of the field more crop marks were observed by the pioneer aerial archaeologist Major G. W. G. Allen, and in 1946 Richard Atkinson made here the first geophysical survey of an archaeological site. Atkinson had written, as a postscript to the first edition of his book *Field Archaeology*, published that year:

> As this book goes to press, the writer is investigating a method of detecting the position of bedrock, and of pits and ditches cut in it, at depths beyond the reach of a probe; it depends upon measurement of the resistance of the soil to the passage of an electric current, and has been used successfully in civil engineering. Any useful results will be published in due course.

A modest beginning to the many words that have been written on the subject since. The results were indeed useful. By the time of the next edition of *Field Archaeology* in 1953, resistivity prospecting had a chapter to itself. The stimulus for its development came from the convergence of two very different pressures at the end of the Second World War: the great harvest of aerial discovery (Wilson, 1982) and the rapid destruction of the riches thus revealed by reconstruction, quickening development and increasingly powerful methods of cultivation. The work at Dorchester was some of the earliest 'rescue' or 'salvage' archaeology. Gravel extraction there was rapidly approaching the features discovered by Allen, a remarkable cluster of monuments of curious forms, including henges. But the field was large, with few visible landmarks, and the photographs oblique, making the features difficult to locate precisely, so that a ground-based method of 'remote sensing' was urgently needed.

Atkinson writes:

I came across an account of resistivity surveying quite by chance in 1946, in a civil engineering journal which had been sent to me by J. J. Leeming, then Deputy County Surveyor for Oxfordshire, which contained an article by him on roads and archaeology. He was a keen amateur archaeologist. In the same number of the journal was an article on resistivity surveying for site exploration for dams, giving the name of Evershed and Vignoles as manufacturers of the appropriate instrument. It seemed to me that the same method might be useful for detecting buried archaeological sites, and so I got in touch with Evershed and Vignoles, who generously did a skeleton survey of the site at Dorchester-on-Thames which I was about to excavate, charging me only the cost of their expenses. The results were promising, but their technique was very slow, because they were using only four equally-spaced electrodes, in the form of hollow copper tubes with a rather bluntly-pointed steel insert at the bottom, which were inserted by repeatedly dropping an iron bar into the tube. It took two whole days to survey four diametric traverses across the site about an assumed centre.

I then hired a Megger from them, and devised the system of five steel-rod electrodes and wafer switch which I have since described, and re-surveyed the site. This is how it all began.

I gave the first public account in a lecture to the Society of Antiquaries on 16 January 1947, the same day that I was admitted Fellow. The then Editor of the *Antiquaries Journal* declined

to publish it, as too technical. In 1951 I was invited to contribute a chapter on resistivity surveying to a French book, *La découverte du passé* (Paris: Picard, 1952), edited by Mlle Annette Laming, who had worked with me on excavations at Dorchester, located by this means. The first British publication was in the second edition of my *Field Archaeology*.

In testing resistivity, Atkinson must also have been influenced by the fact that the soil moisture variations that it responds to are also a major contributor to the creation of the crop marks that reveal such sites as his. He was able to persuade the Ashmolean Museum, for which he worked, to purchase a Megger Earth Tester. The Museum also houses the archive of Major Allen's air photographs, and in 1948 staged the first exhibition of these. By one of those chances that shape lives, I was working nearby for the Inter-Service Photographic Interpretation School, RAF Nuneham Park, and was greatly impressed by both the exhibition and the resistivity work at Dorchester, only three miles away. At Cambridge, the post of Curator in Aerial Photography was newly established. Remote sensing in archaeology was on the move.

The Megger Earth Tester in those days was built in the venerable tradition of mahogany and

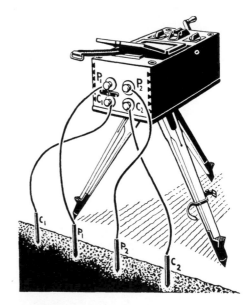

Fig. 1. Diagram of Megger Earth Tester arranged for Wenner resistivity measurement.

brass (Fig. 1). It was powered by a hand-driven generator producing a high enough voltage to make handlers of uninsulated probes jump, and the method of operation was a fine example of electromechanical technology. The Megger was well established for soil studies in civil engineering, in which it was mainly used at fixed

Fig. 2. Contour plan of one of the first resistivity surveys of a site at Dorchester-on-Thames. Values in ohm-feet. The actual ditches as excavated are shown by stippling. After Atkinson (1963).

stations where the spacing between the four necessary probes was varied for soil testing in depth. The basic requirement of archaeological prospecting is different: rows of readings are taken at constant probe spacing on the assumption that undisturbed ground will give fairly uniform values, while archaeological features will cause anomalous readings (simply known as 'anomalies') that will stand out when the survey is plotted. Atkinson worked out a novel, rapid method of survey called the 'leapfrog' and the

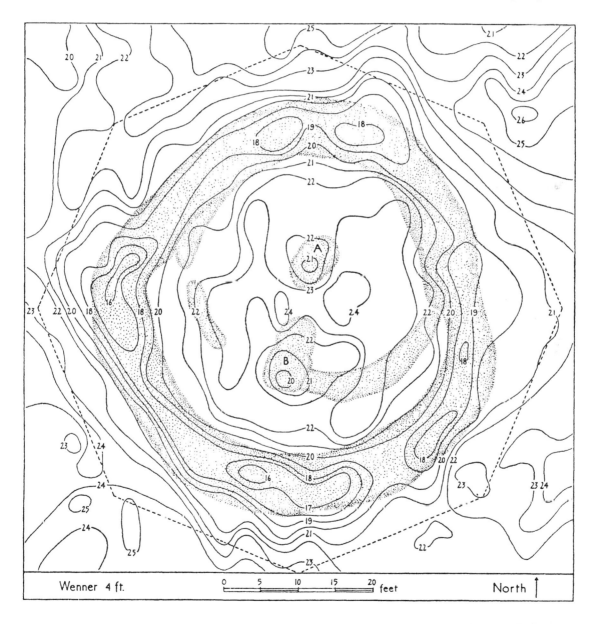

switching system to operate it, which he alludes to above and will be described in Chapter 2.

The simplest way to represent lines of readings is by graphs, but Atkinson readily appreciated the value of area surveys with their potential for revealing man-made patterns in plan. The Dorchester survey shown here was a good example of this early work (Fig. 2).

The Dorchester sites, consisting of moist, silted-up features cut into well-drained gravel, gave encouragingly clear resistivity responses. For 12 years from 1946, when Atkinson initiated the technique, resistivity, mainly with the Megger, remained the sole instrumental method for detecting and planning archaeological sites, although subsequent success was variable. In the 1950s, technology advanced: the transistor was available, and allowed the development of circuits which could be complex, yet compact, highly reliable and low in power consumption. This encouraged John Martin and myself in 1956 to develop a resistivity meter with archaeology specifically in mind – the Martin-Clark, a spin-off from our work in the Distillers Company Instrumentation Section during its extraordinary and enlightened first years under the late Charles Munday.

The Roman town of *Cunetio*, in Wiltshire, had recently been discovered from the air by J. K. S. St Joseph, the Curator at Cambridge. Excavations there, directed by Kenneth Annable, first with Ilid Anthony and subsequently myself, provided a test-bed for the new instrument. The first

Fig. 3. (Bottom) Using the prototype Martin-Clark meter at *Cunetio*. Left to right: Richard Sandell, Kenneth Annable and the author. (Top) A problem: trying to revive flat batteries with a candle borrowed from the local pub.

Fig. 4. *Cunetio*, Wiltshire. One of the air photographs taken by J. K. S. St Joseph which revealed the defences of the Roman town, looking east. The arrows point along the crop marks of the east wall and the south wall with two visible bastions.

Fig. 5. (Right) The *Cunetio* bastion search. (1) and (2) are traverses across the wall foundation to establish its position. On the basis of these, (3) was laid out parallel with its outer face so as to intersect the bastions. The graphs show that the resolution of the meter was 0.2 ohm. A minimal excavation of the bastion giving the larger anomaly – and therefore likely to be the better preserved – is shown in the bottom plan. It was just sufficient to produce the information required: the plan of the bastion, and the fact that it was integral with the wall and not a later addition. Traverses (1) and (2) also show indications of the earlier ditch system visible on the air photograph.

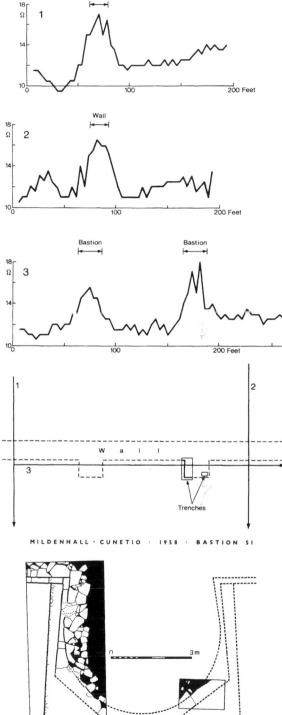

version was a simple two-electrode, transistorized Wheatstone bridge, and even this, in spite of contact resistance problems, was readily able to detect the massive foundations of the defensive wall around the town (Clark, 1957). This led quickly to the development of a more suitable instrument using a basically conventional manual null balance circuit and a newly designed switch that simplified Atkinson's leapfrog probe-moving method (Fig. 3). The null balance of the circuit was initially detected audibly, using earphones, but because of problems with subjectivity in judging a rather imperfect null, this was replaced by a dial.

The value of archaeological geophysics was brought home to me when I arrived late at the site after a struggle with some instrumental problem. On the basis of estimates from the oblique air photographs, the archaeological team had cut a substantial trial trench in search of the town wall, which subsequent resistivity measurements showed to be about 7 m (23 ft) away. The instrument was used to trace the whole circuit of this wall, and the search for the bastions provided a fine and convincing example of the economy of effort achievable by a geophysical technique (Figs. 4, 5). This prototype is now in the Science Museum.

Meanwhile, another momentous development was under way. It had been known since the nineteenth century that fired clay, including pottery, was weakly magnetic. Clays contain iron oxides which are magnetized by the Earth's field as they cool down from high temperatures. By this process, called thermoremanent magnetization, the direction of the geomagnetic field at the time of use is fixed in the kiln. This direction changes over the years and the date is determined by making measurements on samples taken from the kiln, matched against a reference curve of the directional changes in the Earth's field with time.

In 1957, the Canadian physicist John Belshé gave a lecture to the Society of Antiquaries reporting on pioneering work at Cambridge on this method of dating in Britain. In the discussion following the lecture, Graham Webster asked if it would be possible for a magnetometer to detect buried burnt clay, such as a pottery kiln. Belshé said he could see no reason why it should not; in fact, he had demonstrated the previous year that a kiln would produce an appreciable magnetic signal. Webster had good reason to ask this question. The A1 Great North Road was being improved, and there was evidence that there could be pottery kilns somewhere along the 3-km (2-mile) stretch to be re-routed and widened past the Roman town of *Durobrivae*, at Water Newton near Peterborough. He was in overall charge of the daunting project of searching this lengthy strip of ground for kilns, until then using bull-

dozers, which churned up the ground so much that nothing could be seen. Immediately after the lecture, Webster contacted the Department of Geology at his own University of Birmingham and Martin Aitken of the newly established Research Laboratory for Archaeology and the History of Art at Oxford. Aitken and Edward Hall, the head of the Laboratory, realized that this could be a suitable application for a magnetometer based on the proton free precession principle recently discovered by Packard and Varian (1954) and developed for field use by Waters and Francis (1958).

Aitken and Hall had only eight weeks before the Water Newton work was scheduled to recommence but, after many difficulties, their prototype instrument was ready on time and first operated in the field in March, 1958, 'to the accompaniment of a light fall of snow' (Fig. 7). It had a sensitivity down to about 1 gamma, or

Fig. 6. A section of the *Cunetio* town wall.

1 nanotesla (nT) in the new SI nomenclature – about 0.002 per cent of the Earth's field strength in Britain – yet was fast and tolerant of rough treatment. A conventional Askania torsion-fibre instrument was tested by Tony Rees of Birmingham, but its slow speed of operation – one reading in five minutes, as opposed to five seconds for the proton instrument – made its use impractical in the archaeological context (Aitken *et al*, 1958).

In the course of seven days 5 ha (12 acres) were covered. Pipes and bedsteads were found but only one kiln, on the last day. Great were the celebrations in the Haycock Inn at Wansford that night, and the excitement, even wonder, of that moment was captured by Aitken in a subsequent BBC broadcast (Aitken, 1959):

> Last spring, in the fields bordering the Roman town of Durobrivae, near Peterborough in Northamptonshire, I measured the speed of gyration of the nucleus of the hydrogen atom; that is, a proton. I measured this speed at five-foot intervals over a large area, and in one spot I found that it was a little faster than elsewhere. A test-hole was dug and at a yard down we found the upper rim of a Romano-British pottery kiln...

The instrument also proved to be sensitive to soil-filled features such as pits and ditches. This discovery was first regarded as a mild irritation by the scientists, but as a considerable bonus by the archaeologists. The response was a surprise, but accepted at the time as probably due to the large amount of pottery in the pits. But Le Borgne (1955) had recently shown that topsoil is normally of greater magnetic susceptibility than subsoil, so that pits silted with topsoil might be expected to be detectable. However, the anomalies at Water Newton reached the astonishing level of about 60 nT, clearly (we now realize) because the pits were associated with a great deal of burning, the most potent factor in soil susceptibility enhancement, as it is in the different phenomenon of thermoremanence.

Thus it seemed probable that magnetic detection would be especially sensitive to many types of feature associated with human occupation with its concomitant fires, and this has been amply proved to be the case. Magnetic survey with the new magnetometer was immediately established as a powerful technique in archaeological prospecting: fast and simple to use because it requires no probe insertion, and above all because it is actually selective of the effects of human activity.

With two powerful techniques established, the pace quickened. A potentially important centre, the Archaeological Prospecting Section of the Lerici Foundation (Sezione Prospezioni Archeologiche della Fondazione C. M. Lerici del Politecnico di Milano), had been established in 1954 and was initially much concerned with the detection by resistivity and subsequent rescue of Etruscan tombs under threat from the notorious *tombaroli* tomb robbers. A much-publicized Lerici procedure was first to detect a tomb, then to drill a small hole through its roof and inspect the interior by means of a periscope to assess the quality of the grave goods and whether there were any frescoes on the walls (Bacon, 1960; Deuel, 1969; Lerici *et al*, 1958). Guided by Richard Linington, the Lerici Foundation was to become an important centre of archaeological prospecting in Europe, providing surveys in a number of

Fig. 7. Martin Aitken using the prototype proton magnetometer. The detector bottle is supported on a tripod in the background.

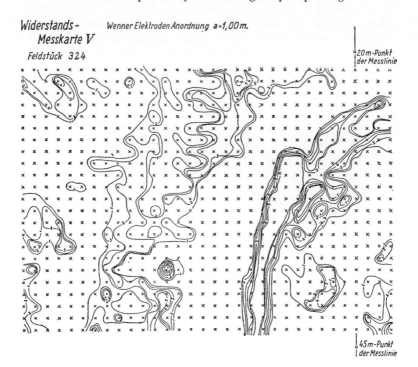

Widerstands-
Messkarte V   Wenner Elektroden Anordnung a=1,00m.

Feldstück 324

○20m-Punkt
der Messlinie

Fig. 8

45m-Punkt
der Messlinie

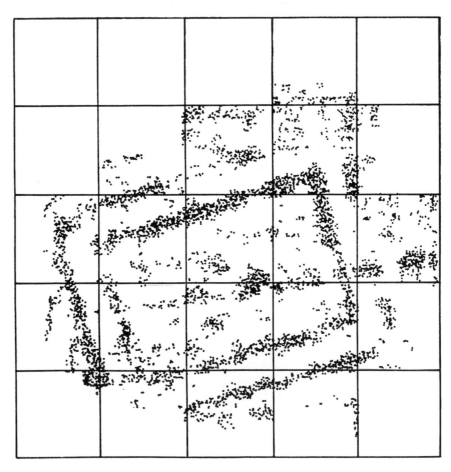

Fig. 9

Fig. 8. The first published resistivity plot using filtering, of a corner of the wall foundation of the Roman *Colonia Ulpia Traiana* at Xanten in Rhineland. Contours clearly outline the wall with a possible bastion. The ditch, separated from the wall by a berm, is less clearly defined, probably in part because the filter was less appropriate to its width (Scollar, 1959).

Fig. 9. One of the first examples of a computer-generated filtered dot-density plot. A magnetic survey of a Roman ditched enclosure in the Rhineland. The squares are 20 m (66 ft) (Scollar, 1966).

countries in the East and West. From 1964 the Foundation ran instructional courses in Rome, and in 1966 it established the journal *Prospezioni Archeologiche*.

In 1958, Irwin Scollar, at the Rheinisches Landesmuseum in Bonn, was pioneering the computer processing of resistivity data, including the use of spatial filtering (Fig. 8). From the early 1960s he was designing his own magnetometers and collecting data automatically onto punched tape. He was an advocate of the use of differential magnetometer measurements with a fixed and movable sensor to remove geomagnetic background variations, and devised the subsequently much-used 'dot density' method of representation (Fig. 9). Scollar's data-logging system filled most of the available space in a Volkswagen minibus, but such has been the subsequent march of electronic development that a modern laptop computer can serve as a miniature data logger of similar power to this pioneering system, with the added ability to provide a picture of the survey as it progresses.

By 1960, Martin-Clark resistivity meters were available commercially, as were two versions of proton magnetometer, directly descended from the first instrument used at Water Newton and manufactured by the Littlemore Scientific Engineering Company in association with the Oxford Laboratory. The more accessibly priced and popular of these was the 'Maxbleep', an instrument with two detectors arranged in a gradiometer configuration to reduce the effect of external interference and remove background geomagnetic variations. By means of relatively simple circuitry, signals from a detector close to the ground and a reference detector at a higher level were mixed and put out in audio (as well as visual) form, any difference in signal being represented by beats increasing in frequency with the signal difference (Fig. 10). These beats produced a bleeping sound which inspired the instrument's name. The electronics of these magnetometers were separated from the detectors at the end of a long cable, to avoid problems with magnetic components. A smaller 'Minbleep' for one-man operation, in which such magnetic interference was minimized, was never fully successful.

Archaeometry, the journal of the Oxford Laboratory, chronicles some specially significant developments in Volume 7, for 1964. Beth Ralph, of the University of Pennsylvania Applied Science Center, introduces the first archaeological applications of an instrument using the principle of optical pumping, the rubidium magnetometer of Varian Associates, potentially capable of 0.01 nT resolution and a gain of a hundred times over the sensitivity of the proton instrument at that time. An exacting test for this was the search for the site of the ancient Greek city of Sybaris, in south Italy, where buildings of non-magnetic stone lay beneath slightly magnetic alluvial clay that had overwhelmed the remains. Although the proton magnetometer had been able to locate later Roman walls, more deeply buried Greek walls were not expected to give anomalies greater than about 2 nT, at the limit of the capability of the absolute instrument used at that time. The first use of the new and sophisticated instrument was beset with an ample selection of the kinds of technical problems all innovators recognize with sympathy, but later improved equipment was triumphantly successful in finally locating the elusive Sybaris.

The second highly significant development was a fluxgate gradiometer, built and described by John Alldred of the Oxford Laboratory. The electronics were in a case separate from the detectors, which were mounted 1.22 m (4 ft) apart in a double tube carried vertically by a kind of yoke fixed close to its top, and carried on the shoulders of two people, in front and behind. A third detector, mounted 30 cm (1 ft) above the lower detector, gave the option of a shorter baseline for testing whether anomalies were caused by deep or superficial objects – a sensible feature that has never been followed up. The objective in building this instrument was to achieve greater speed than existing proton equipment, and it was found to be most effective when carried continu-

Fig. 10. Maxbleep in use by the Surrey Archaeological Society at Leatherhead. This particular instrument was, in fact, something of a hybrid, incorporating elements of the prototype Minbleep.

ously along lines 1.52 m (5 ft) apart, and noting the position of any indicated anomalies. The best speed achieved was twelve 15 m (50 ft) squares in an hour, which is comparable with modern methods, although this rate of coverage could only be achieved in an area containing few anomalies requiring investigation.

In 1967 John Musty established the Geophysics Section of the Ancient Monuments Laboratory (AML), part of the Inspectorate of Ancient Monuments of the Ministry of Public Building and Works, which also retained a Test Branch at Cardington, Bedford. For some years, any archaeological prospecting required had been undertaken by this organization, which had developed an innovative multi-probe resistivity system, operated manually and based on the high quality manual balance Tellohm resistivity meter.

The AML Geophysics Section at first consisted only of myself, so I had a strong incentive to use equipment operable by one person. The square resistivity array was such a system, and much use was made of it in the first years. It was notably successful at Burton Fleming on the Yorkshire Wolds (now in North Humberside) for tracing the ditches of the Late Iron Age square barrows for which the area is famous, and which were being investigated by Ian Stead (Chapter 7). A survey of the same site with the Wenner resistivity configuration at 1.22 m (4 ft) spacing was unsuccessful, as was a survey with a proton magnetometer, except that a substantial response around the centre of one of the barrows suggested that it contained remains of a chariot with iron fittings, as a few barrows of this type were known to do. By 1968, when this survey was made, a number of new techniques were being applied in archaeological prospecting, and this site, to which the established ones reacted so variably, seemed a suitable test-bed for comprehensive comparisons – and there was an excellent

Fig. 11. (Above) The Burton Fleming experiment. Michael Tite and Christopher Mullins using a pulsed induction instrument on a grid formed from a 50 ft (15 m) net of plastic-coated clothes line with a 10 ft (3 m) mesh. The tower carrying the EMI infra-red experiment is in the background.

Fig. 12. (Below) The Burton Fleming experiment. Eric Foster and John Alldred surveying with a portable pulsed induction meter on a 25 ft (7.5 m) grid net with 5 ft (1.5 m) mesh.

inn, the Bosville Arms, to stay at in nearby Rudston.

Invitations were sent out; there was a good response, and all the emerging techniques were represented (Figs. 11, 12). These included two versions of the pulsed induction meter, or PIM, an invention of C. Colani recently developed at the Oxford Laboratory (Michael Tite, Christopher Mullins, John Alldred and Eric Foster); the SCM soil conductivity meter, nicknamed the 'Banjo' and well known for its success at Cadbury Castle (probably Camelot) in Somerset (Mark Howell); and an infra-red detector of the type used in thermal mapping from the air, provided by EMI and set up on a photographic tower (John Gurdler). The working of the PIM and SCM will be discussed in Chapter 4.

In the event, none of these techniques obtained a positive response from the small barrow ditches at Burton Fleming, but the need to explain this was a valuable starting point for developing a fuller understanding of their potential and limitations.

Perhaps most disconcerting was the inability of the PIM, designed primarily as a high sensitivity metal detector, to see anything of the massive magnetic anomaly within one of the barrows. In fact, when the site was excavated, its cause was found to be a most unexpected component of igneous material in the chalk gravel filling of the shallow valley in which the site lay, which must have been transported by glacial action. Thus the negative PIM readings were evidence that the anomaly was not caused by iron, as the use of the magnetometer alone suggested.

The SCM was expected to detect the low resistivity of the ditch fillings as revealed by the square array survey, but again there was no response. Suspecting that the reason might have been that the instrument actually measured magnetic susceptibility rather than conductivity, the test led Tite and Mullins to undertake a detailed research programme on the instrument which showed that this was indeed the case (Tite and Mullins, 1970). Its performance here and on other more suitable sites was also often disappointing, and this proved to be due to poor penetration. In retrospect, its success at Cadbury Castle was clearly due to almost ideal conditions of shallow soil cover and strong susceptibility contrast. The research on this instrument stimulated Tite and Mullins to extend their studies into the whole subject of soil magnetic susceptibility, and their resultant publications are fundamental references (for example Tite and Mullins, 1971; Tite, 1972a, Mullins, 1977).

The infra-red experiment was operated round the clock for several days in order not to miss any conditions of diurnal heat absorption or emission that might be affected by the thermal conductivity of the buried remains. No convincing response was obtained, but the field had been harrowed, leaving a superficial layer of small earth clods sufficiently insulated from the soil beneath to make the detection of any effect very difficult.

*

An important development in Britain was the establishment of the Bradford University Department of Archaeological Sciences (as it is now called), in association with the Department of Physics. Under the energetic leadership of Arnold Aspinall, this has specialized in providing balanced undergraduate courses in archaeology and archaeological science, and postgraduate science courses for archaeologists. Bradford graduates have spread scientific sense through archaeology to its great benefit. Always strongly involved with prospecting, especially resistivity, Bradford was responsible for the development of the twin electrode method after its invention by Schwarz in Switzerland, and for the automatic Bradphys resistivity meter, the ancestor of the Geoscan RM4. Roger Walker, the founder of Geoscan whose developments have had a profound influence on archaeological prospecting, was himself a product of the Bradford Department.

In 1964, the Oxford Laboratory initiated an annual meeting on a Saturday in spring. This was called the Maxbleep Symposium, at which a few people who had bought these instruments could have any necessary repairs done, and give brief accounts of their experiences in using them. A tour of the Laboratory gave an opportunity to view its current work. By 1968 the meeting had become a one-day Symposium on Archaeological Prospection and included European contributors; from this it developed into the Symposium on Archaeometry and Archaeological Prospection, lasting several days. The high spot was a magnificent dinner in an Oxford college where we were the guests of Edward Hall, Head of the Laboratory. From 1975 the meeting became the International Symposium on Archaeometry and

Archaeological Prospection, travelling the world. This continues today, taking place every other year.

There has been an interesting evolution: prospecting has been left out of the title of some more recent meetings, a sign of a levelling off in development accompanied by a great expansion of analytical and dating methods. Another sign of this was when the journal *Prospezioni Archeologiche* ceased publication in 1974. There has since been a revival of interest, partly because the development of microcomputers has facilitated high quality rapid recording, and new instruments and plotting methods have appeared to match their facility. Electromagnetic instruments have made considerable strides, and ground-penetrating radar is becoming highly effective and more widely used. There are also simply more

professional practitioners as the techniques become more routinely required in archaeology. The word 'Prospection' has reappeared in the title of the International Symposium, but there is still a multitude of other aspects jostling for position. Something akin to the spirit of the earlier Oxford meetings has been rekindled in Britain with the first of a new series of meetings devoted to prospecting at Bradford in 1985. Another landmark was the first session devoted to archaeology at the prestigious annual meeting of the American Society of Exploration Geophysicists, in Atlanta in 1984.

Mention must also be made of the *MASCA Newsletter*, now *MASCA Journal*, published since 1965 by the Applied Science Center for Archaeology of the University Museum, Pennsylvania. Originally issued free internationally, this has provided a valuable record of developments in the USA, including, in the area of prospecting, the earliest work with the highly sensitive optically pumped magnetometers and ground penetrating radar.

*

At the Oxford meeting in 1968, Frank Philpot, then at the Plessey Company, demonstrated a

Fig. 13. The first continuously recorded fluxgate gradiometer survey, of two kilns at Linwood in the New Forest. Re-drawn traces from Plessey report ERL/N 161 U.

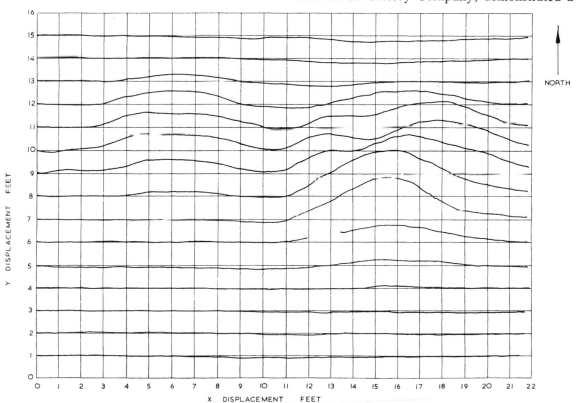

fluxgate gradiometer that was more compact than the first design by Alldred, and easily carried by one person. Another idea of Philpot was to exploit its advantage of continuous output by linking it to an XY plotter, a chart recorder in which separate signals drive the vertical and horizontal movements of the pen, rather than having a continuously driven chart as is used, for example, in a barograph. Before long we found ourselves in a glade at Linwood in the New Forest, with the equipment set up to plot two pottery kilns known to exist from excavations by Heywood Sumner.

The survey was to be made in a series of parallel straight lines, marked out with strings. The gradiometer output was recorded by the vertical movement of the pen, and the distance along the lines by its horizontal movement. This was achieved by means of a string attached to the instrument at one end and wound around a drum linked to a potentiometer at the other, so that movement of the instrument produced a proportional output from the potentiometer to drive the pen. At the end of the traverse along each ground string, the pen was manually stepped up ready for the next, so that the final plot represented a series of very high-resolution magnetic profiles across the buried kilns (Fig. 13). Later, these were translated into a contour plot which gave an excellent detailed representation of the kilns (Fig. 14), confirmed in excavation by Vivien Swan.

The string from the drum to the instrument was kept taut by the remarkable expedient of having an electric motor attempting to drive the drum in opposition to its unwinding. The movement of the instrument, over a very short distance, was awkward and at a snail's pace and – like a snail – it was actually moved along the surface of the ground. The recording equipment was driven by inverters supplied by massive batteries because portable equipment was rare in those days. But a principle, and a great step forward for magnetic surveying, had been established which is recognizable in methods used to this day. The fluxgate revolution had begun.

David Haddon-Reece and I set about adapting this principle to the speed of operation required in routine survey (Clark and Haddon-Reece, 1973). A potentiometer 30 m (98 ft) long was constructed by stretching a resistance wire between two tripods. The instrument was carried along-side this and at the same time a contact 'trolley' was moved along the wire, recording the instrument's progress accurately on the chart. The trolley also tripped switches at each end, automatically stepping up the recorder chart pen by the right amount for the next traverse by means of a stepping potentiometer operated by an ex-telephone exchange uniselector. Two traverses were made for each position of the wire by carrying the instrument first along one side of it and then the other. The wire gave much trouble because of bad contact with the trolley due to kinking and corrosion and, amazingly, spiders leaving trails of webs along it overnight – quite apart from a tendency for cows to walk through it, dragging the whole assembly around the field. It was therefore quite soon abandoned in favour of a sealed rotary potentiometer linked by Meccano gears from the nearest toy shop (what would we do now that these are no longer made?) operated by a pulley and string, which could be readily demounted (Fig. 54). This was less precise because of a tendency to slipping and stretching, but these problems were eventually largely solved by the use of nylon monofilament for the string, and the liberal use of rubber bands on a flat pulley wheel.

Crucial to the success of this system was the excellent little XY plotter known as the Servoscribe M, almost the only one available powered by small dry batteries. Surveys were made in blocks of 15 traverses and assembled as montages. This was normally done in the laboratory where they were tidied up and photocopied for report production; or a montage could be prepared in the field if it was needed as an immediate guide to excavation planning. The immediacy of trace recording also made it easy to see where archaeological anomalies ceased so that time was not wasted on unproductive areas, and its detail made possible the recognition of the characteristic signatures of types of archaeological feature, as well as the unambiguous identification of signals caused by modern iron rubbish.

Many dramatic survey plans were produced with this system but, in spite of its enormous advantages, it was inflexible. The presentation was restricted to traces and the recording sensitivity had to be decided at the beginning of the survey – even though it might subsequently prove to be inappropriate. The attainable sensitivity was limited to 1 nT at best because of various limitations in the instruments and the recording system. Some surveys were converted to dot density and contour form by means of manual trace-follower digitizers, but this was excessively laborious. Digital data logging in parallel with the trace recording was tried but the loggers available were not well adapted to this work, rather to the collection of readings at regular and quite widely spaced intervals of time, as at weather stations.

The appearance on the market of the Epson HX20 portable microcomputer in the early 1980s stimulated major development. Here at last was a battery-powered instrument that could be used for flexible data logging, while retaining the immediacy of trace plotting by being able to produce a small plot of the survey at the same time. The Ancient Monuments Laboratory has now replaced this pioneer computer with a Toshiba portable with a hard data storage disc, higher processing speed and a screen large enough to display traces on the same scale as those plotted on paper by the Servoscribe M. Used with other

Fig. 15. The Linwood kilns after excavation. The further one is built in the stoke-hole of the first.

Fig. 14. (Below) The Linwood survey converted to contours and superimposed on a plan of the kilns as excavated by Vivien Swan.

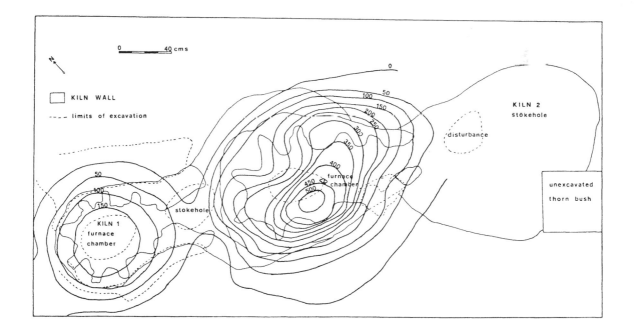

modern equipment, as we shall see, this brings speed of data gathering, efficiency and effectiveness to a level that was hard to foresee a few years ago. An honourable mention must be made of a survey system devised by Sowerbutts and Mason (1984). A precursor of those with miniature computers, it made use of an Apple desk-top microcomputer mounted on a custom-built wheelbarrow and powered by a car battery with an inverter. Another innovation was an ultrasonic distance transducer which required no strings, simply a transmitter on the instrument and a receiver fixed on the ground. It was used, in combination with excellent display software, principally for solving geotechnical problems such as the location of old mine shafts.

*

This chapter has said little of developments abroad, many of which have been chronicled in the pages of *Archaeometry*, *Prospezioni Archeologiche*, the *MASCA Newsletter* and in the proceedings of successive international Archaeometry conferences. Much crucially important underpinning theory, especially in the area of the processing and presentation of survey measurements, has been provided in papers by Irwin Scollar and by the late Richard Linington of the Fondazione Lerici. A distinguished team at the Centre de Recherches Geophysiques, at Garchy in France, notably Albert Hesse and Alain Tabbagh, has tended to specialize in evaluating and developing the potential of techniques not yet in general use, for instance thermal detection and the utilization of very low frequency radio signals and other electromagnetic techniques. They have also given substantial attention to resistivity, for which they have developed high-speed tractor-drawn systems in contrast with the British preference for lightweight, highly portable equipment. John Weymouth at the University of Nebraska has been the focus of much of the activity in the USA (Weymouth, 1986).

*

Even as these words were being written, significant developments have taken place. The wide availability of high quality half-tone printers has brought unprecedented refinement to survey plotting, making visible subtle features that our instruments have been detecting but we have previously been unable to display in sufficiently comprehensible form. Improvements have been made in ground-penetrating radar. One might be forgiven for thinking that the peak of development in archaeological prospecting had been reached. Experience teaches that it has not.

Chapter 2
Resistivity

The electrical resistance of the ground is almost entirely dependent upon the amount and distribution of moisture within it. Buried remains affect this distribution, and can be detected with instruments. Stone, for instance, is generally more moisture-resistant than a clay subsoil or the soil filling of a ditch, while soils will hold differing proportions of water depending on their texture. Clay and soil may have resistivities of 1–10 ohm-metres (Ω-m), and porous rocks 100–1,000 Ω-m, while non-porous rocks will rise to anything between 10^3 and 10^6 Ω-m. These differences may be distinguished by measurements of the resistivity of the ground, enabling archaeological remains to be discovered and planned.

When one first encounters a soil resistivity survey in progress, two things often seem most puzzling. How can objects deep under the ground be detected by electrodes inserted only a few centimetres? And why are four electrode probes (sometimes even five) needed? I will start with these and some other fundamentals.

Principles

Definitions, units and relationships An electric current is caused by the movement of charged particles. In metals these are mainly electrons; in liquids they are molecules which have separated into their positively and negatively charged constituent parts, known as ions. When an electrical potential difference, or voltage, is applied between the ends of an electrical conductor (a wire for instance), a current flows through it, the size of the current depending upon the resistance of the conductor. The symbol for current is I, which is measured in amperes, symbol A, often shortened to amps. The symbol for resistance is R, measured in ohms (often represented by the Greek letter omega, Ω). The symbol for potential difference, or volts, is V.

Taking the flow of water in a pipe as an analogy, potential difference is the water pressure, and current is the flow. The narrower the pipe, the higher the resistance, and the lower the flow for any particular pressure. Thus, current is inversely proportional to resistance:

$$I = \frac{V}{R} \qquad \text{i.e.} \qquad R = \frac{V}{I}$$

This means that resistance is the ratio of potential difference to current flow. Rearranging the formula again, $V = I \times R$, which means that when a current flows through a resistance, a potential difference appears between its two ends. *Resistivity* is *specific resistance*, which enables the resistances of different materials to be compared in a standardized way. In the modern International System of units (SI), the resistivity unit is the ohm-metre (Ω-m): the resistance of a 1 metre cube of the material when a potential difference of 1 volt is applied between two opposite faces of the cube. The symbol for resistivity is the Greek letter rho (ρ). Materials used as electrical conductors, such as copper, have a very low resistivity but other materials and metals can have high resistivities, to the point where they become insulators.

Resistivity and the soil In ground measurements, very little current is carried by the soil and rock themselves, most of which are insulators. Rainfall, which contains dissolved carbon dioxide and carbonic acid from the atmosphere, forms conducting electrolytes by reaction with the minerals in the soil, which also contains weakly conductive organic acids. All these break down (dissociate) into positive and negative ions which are the actual carriers or conductors of the electricity. For instance, on a chalk soil:

$$CaCO_3 + CO_2 + H_2O = Ca^{2+} + 2HCO_3^-$$

This means that chalk (calcium carbonate, $CaCO_3$) plus carbon dioxide plus water react to form calcium bicarbonate, $Ca(HCO_3)_2$, which dissociates into a calcium ion with two positive charges and two negatively charged carbonate ions. Humic acids also make a contribution to conduction in soil.

Soil resistance measurements are made between electrodes or probes (terms used interchangeably) pushed only sufficiently far into the ground to make adequate contact, usually a few centimetres. Being conditioned by the everyday image of conduction through wires, one's first thought is that the electric current will pass close to the surface, straight from one electrode to the other. In fact, the Earth is for practical purposes defined as a 'semi-infinite medium', which means that it has just one boundary, the surface. The current is thus unconfined and, in its search for easy paths and also because similarly charged ions repel one another, spreads deeply into the ground.

It is easy enough to measure the resistance of a length of wire. You connect a simple instrument containing a battery and a current meter

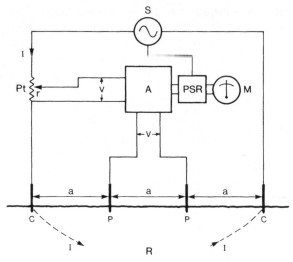

Fig. 17. A practical earth resistivity measuring circuit, as used in the Martin-Clark meters. Contact resistance problems are overcome by using separate potential and current electrodes, and polarization by the use of an AC supply. The Wenner electrode arrangement, with equal spacing, a, is shown. The working of the circuit is described in the text.

(ammeter) to its two ends and measure the current that flows. The meter is calibrated in terms of resistance. A similar measurement on the ground, however, presents problems (Fig. 16). The electrodes have a very small contact area in relation to the overall volume of ground traversed by the current; and, compared with metal, the soil is a poor conductor, especially near its surface where it tends to be relatively dry. In combination, these effects create much higher resistance immediately around the probes than is encountered by the current in the deeper ground that one wants to measure. Variations in this high resistance are inevitable, especially where there are difficulties in inserting the probes. If only two probes were used, these contact resistance variations would swamp significant changes which lie deeper.

The contact resistance problem is overcome by making a four-terminal resistance measurement (Fig. 17). A current is passed through electrodes C, C, and the resulting potential gradient in the ground is sampled between potential electrodes P, P. If the measuring circuit is designed with very high impedance, so that it takes no current from the potential electrodes, then this potential difference divided by the current gives a resistance

Fig. 16. A simple ground resistance measurement. Problems of contact resistance and polarization are indicated symbolically.

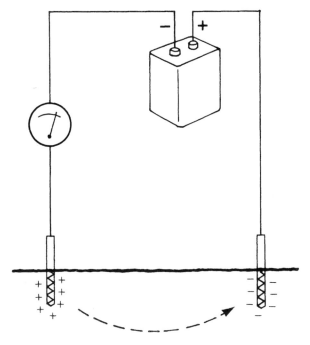

value proportional to the average resistivity of the ground. If contact conditions are bad and the current drops, then the potential measurement also drops, but their ratio – the resistance – stays the same. This type of measurement is also used for low resistivity samples in the laboratory.

Another basic requirement is that the direct current (DC) supplied by a battery must be converted to an alternating (AC) supply to counteract the phenomenon of polarization. Complex electrochemical reactions are involved, but the effect is that, with DC, the current-carrying ions with positive charge are attracted to the negative electrode, where they tend to accumulate, and the negative ions to the positive electrode. Because similarly charged ions repel one another, a barrier to conduction is built up around each probe, and the apparent resistance

rises with time. The use of AC, continuously changing direction, prevents the building up of such barriers and it has other advantages: it enables the circuit to ignore earth currents from the mains electricity supply, as well as any natural telluric currents in the ground and electrochemical currents that may be generated by interaction between the probes and soil electrolytes causing a battery effect. The essentials of a practical circuit incorporating these principles are shown in Fig. 17 and discussed below.

The real function of the current electrodes is to set up a field of potential gradient in the ground which is sampled by the potential electrodes. As in Fig. 18 (top), this can be represented in cross-section by contours of equal potential – equipotential lines – which are at right angles to the direction of current flow. Fig. 18 (top) also shows

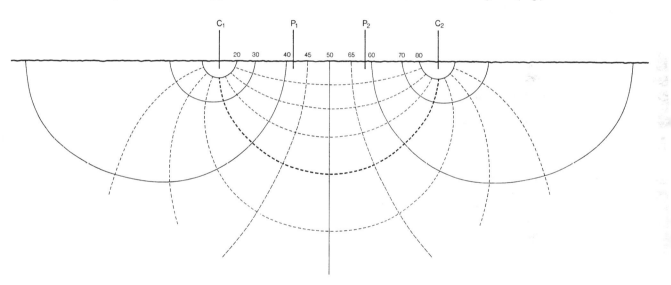

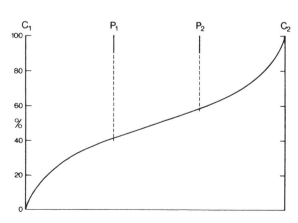

Fig. 18. (Above) Measurements on conductive paper using a Servomex Field Plotter, simulating current and potential distribution in a cross-sectional slice through the ground in the plane of a line of electrodes arranged in the Wenner configuration. Equipotential lines are marked with the percentage of the total potential difference they represent. Current flow is indicated by the broken lines crossing these. Note that half the current radiating from electrodes C_1 and C_2 flows above the heavy semi-circle with maximum depth of half their separation: this is the most sensitive region for detection. (Left) Plot of the potential gradient between C_1 and C_2.

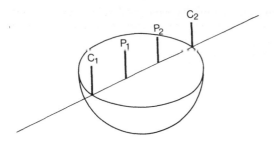

Fig. 19. The hemisphere of detection for the Wenner configuration.

that in a homogeneous soil half the current spreads to a depth of half the spacing between the current probes. The average, or apparent, resistivity to this depth is obtained by multiplying the readings obtained by an appropriate factor. As it spreads sideways as well as downward, the current sweeps through a substantial cross-section of ground, even with quite modest probe spacings (Fig. 19).

If topsoil depth and lithology are uniform, the pattern of equipotential and current flow lines will not vary, and lines of readings across a site will be similar; but buried archaeological features are likely to cause anomalous readings that show up against the general background level, as we shall see below.

Instruments

Instruments can be broadly divided into manual balance and automatic types. Manual balance instruments are designed to give a resistance reading by dividing the voltage difference between the potential electrodes by the current flowing through the current electrodes, while the automatic instruments designed to date use a different approach to circumvent this. The design objective is the fastest possible acquisition of accurate measurements at a fixed electrode spacing. The relative merits of the different electrode configurations mentioned will be elaborated later in this chapter.

Manual balance instruments Most of these are based on the elegantly simple circuit principle shown in Fig. 17. Pt is a potentiometer, a resistance wire with a movable contact which enables a variable proportion of the voltage between its ends to be sampled. Current, I, from the AC source, S, passes through the potentiometer, and through the ground via electrodes C, C. Electrodes P, P sample the potential gradient created in the ground by I, and the voltage, V, between them is applied to the high input impedance amplifier, A. The voltage between one end of Pt and its moving contact is also applied to A, but in antiphase (i.e. reversed) to the electrode voltage so that one voltage is subtracted from the other. The net output of A is shown by the meter, M, and the moving contact of Pt is adjusted by hand until the meter shows no output. When this balance is reached, the voltage between the fixed and moving contacts of Pt is equal to V. As the same current, I, is flowing through Pt and the ground, the resistance, R, between these contacts (read from a dial linked to the moving contact) is the same as the ground resistance. An important refinement is PSR, a phase-sensitive rectifier, which rejects interference by letting through only signals that are synchronous with the power source oscillator. This is particularly important with the compact, low-power instruments that tend to be used in archaeology, in which the level of interference can be quite a large percentage of the true signal. The frequency of the power source is usually between 67 and 137.5 Hertz (oscillations per second). It needs to be carefully chosen to avoid the mains frequency and its harmonics; not so low that polarization can affect the readings, nor so high that skin effects can reduce penetration.

Manual balance instruments are relatively slow to operate and demand concentration by the operator, but their prices are normally the lowest of any instrument used in archaeological prospecting. They are therefore especially favoured by amateur groups who tend to be well supplied with manpower and time, but less so with money, in relation to the scale of their projects. There are a number of manual balance instruments on the market, including a modern successor to the original Megger. These function well, but only one – the Martin-Clark – has been developed with archaeology specifically in mind.

The Martin-Clark meter is small enough to be hand-held, and runs on dry batteries (Fig. 20). Its most distinctive unique feature is a rotary turret switch from which the leads emerge, designed for the 'leapfrog' method of working. This method was conceived by Atkinson, when he first used

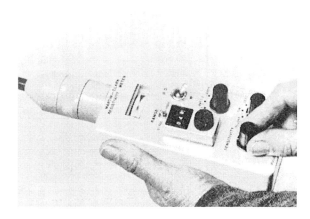

Fig. 20. A modern Martin-Clark resistivity meter, manufactured by the Department of Physics, University of Surrey. The rotary leapfrog switch is on the left. The balancing potentiometer has ten turns and a digital dial, giving good resolution with two ranges of 0–100 and 0–1000 ohms.

resistivity, as a rapid way of making successive readings with a row of four probes in line. It is shown in principle in Fig. 21, and in action in Fig. 22. Without it, all four probes, or their connections, have to be moved for each reading; with it, using five probes, successive readings can be made by moving only one, which is done while a reading is being taken, so that the instrument operator can take readings without interruption.

Fig. 21. (Below) Using the Martin-Clark leapfrog switch, as described in the text. (Inset) The standard electrode probe, made from 1 cm (³/₈ in.) diameter mild steel. Penetration is 18 cm (7 in.) and the offsets are 10 cm (4 in.) wide. The lower offset serves for pushing in the probe by foot and as a depth stop and the upper serves as a handle, bent outward to give clearance for the leg. A hole is drilled at its end to take a banana plug on the lead, which is tied to the probe to take the strain. The lower offset must be placed clear of the tape to avoid disturbing it, and at the right angles to it to avoid uncertainty of effective spacing.

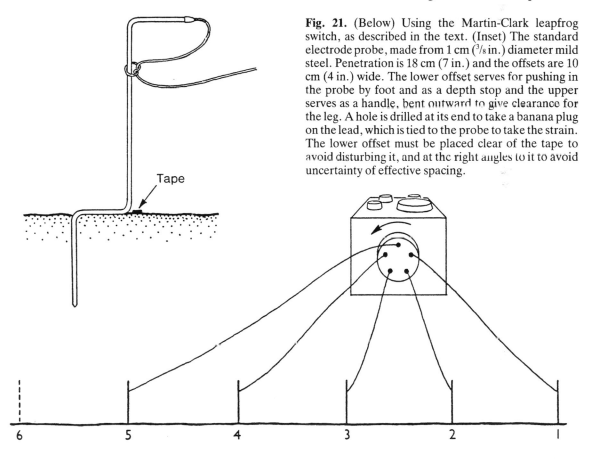

Tape

Resistivity

Fig. 22. (Above) Leapfrog survey.

Fig. 23. (Below) Circuit for leapfrog switch based on wafer components.

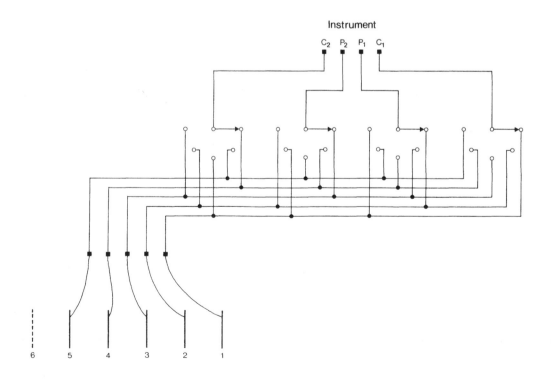

Atkinson designed a leapfrog switch using standard wafer switch components (Fig. 23). The Martin-Clark switch (Fig. 21) is a great deal simpler, but it does have to be specially fabricated. Leads from five electrodes are connected to sockets in a rotatable turret, which engages them successively with four contacts inside the instrument. The active positions are the four lower ones, while the uppermost makes no contact and is called the spare. When the Wenner configuration is being used, the two upper connections are current, and the two lower potential; when the instrument is switched to double dipole (Fig. 33) the two current connections are on one side, and the two potentials on the other.

Setting up in the field is straightforward because the connections to the probes are in the same order as the four live connections to the turret switch, as viewed by the operator from above. The switch is rotated in the direction of travel, which is readily comprehensible to the operator and also counteracts the tendency of probe movement to twist the leads together. However, it does not counteract twisting of the individual leads; therefore, when traversing only in one direction, it is a good idea to turn the probes over once in the forward direction each time they are moved. If a succession of traverses is being made in opposite directions, this is not necessary.

In operation, a reading is taken using probes 1, 2, 3 and 4, with 5 connected to the spare position on the switch. The switch is then rotated one-fifth of a turn to activate probes 2, 3, 4 and 5, and a reading taken while probe 1, which has become spare, is moved to position 6 in readiness for the third reading, and so on.

Colour coding is vital in preventing mistakes with leapfrog surveys. With the Martin-Clark meter, the leads and switch positions are colour coded. With the wafer switch system, probes or leads are coded and each switch position is marked with a row of coloured dots indicating the four active probes in order. Another feature of the Martin-Clark is a switch for changing the probe connections from Wenner to double dipole, so that surveys with both of these configurations can be made simultaneously. The instrument also has built-in battery and lead continuity test facilities. The Martin-Clark can be adapted to twin electrode work by removing the lead assembly and using banana plugs to connect the leads to the appropriate sockets of the rotary switch, which is left in a fixed position. Two 50 m (164 ft) reels of 'figure of eight' domestic electric flex make suitable leads.

Automatic instruments Although the Megger used by Atkinson had a hand-turned generator, its actual readings were automatic. These were achieved with a most ingeniously designed moving coil meter, which effectively performed the V/I calculation by means of separate coils connected to the current and potential electrodes. The disadvantage of this was that an appreciable current was drawn from the potential probes, making the Megger more contact resistance sensitive than modern instruments with high input impedance circuitry.

Modern automatic instruments make use of the constant current principle. The AC current source is designed to maintain a constant current in the ground by varying its voltage in response to changes in the resistance encountered over a wide range. In this case:

$$R = \frac{V}{\text{constant}} \quad \text{i.e.} \quad R \propto V$$

All that needs to be done is to read V, the voltage difference between the potential probes, previously calibrated in terms of resistance.

A popular constant current instrument is the Geoscan RM4, also designed primarily for archaeology. This has a liquid crystal display readout, although a purpose-built automatic data logger is available which greatly speeds its operation. It runs on rechargeable batteries, with a low battery voltage warning. A special twin electrode frame has been designed for the instrument, and a contact resistance indication guides the operator as to how deeply the probes need to be inserted, which in good contact conditions can save the expenditure of much effort in over-deep insertion. In addition, erroneous readings caused by excessive contact resistance are rejected by the circuit and not displayed. A total contact resistance of 40 kΩ (40,000 ohms) between the current probes reduces the current flow by only 0.8 per cent, and therefore readings by about the same amount. This is similar to the contact resistance tolerance of the Martin-Clark meter, and more than ten times better than the original Megger. This impressive figure means that both instruments can cope with dry topsoil conditions, although they have been defeated by the Omani desert fringe, and even in Britain in conditions of extreme drought. Difficulties have also been experienced on glacial gravel and sands in more normal dry summer conditions, notably at Sutton Hoo and other sites in Suffolk, especially when there is the additional drying effect of wind. These problems can often be reduced by the use of the more deeply penetrating individually inserted probes of the Martin-Clark type (Fig. 21) rather than a twin electrode frame.

To achieve their small size (and lack of shocks for the probe handler), the Martin-Clark and RM4 both operate from low-voltage batteries. Their circuits have consequently had to be designed with great care to avoid the effects of stray earthing currents and other mains interference in the ground. To assist in this, the RM4 has a special, somewhat reduced speed, mode of operation for use in urban areas with heavy interference.

Integrated circuits now exist that will make it possible to design instruments that will automatically perform the V/I calculation. These should be capable of an even wider range of contact resistance tolerance than existing instruments.

An increasingly effective class of instruments is the electromagnetic type, which can make conductivity measurements when simply carried over the ground like a magnetometer or placed on the surface. They may ultimately prove to be the most significant development in this type of detection, although it is uncertain whether they will be able to rival the subtlety that can be achieved with conventional resistance instruments by the use of different probe configurations and spacings. The unit of conductance is the siemens, formerly mho, and is simply the inverse or reciprocal of resistance ($S = 1/R$), and the unit of conductivity, σ (sigma), is the inverse of resistivity ($1/\rho$), so the two types of survey are directly comparable. Instruments which have been found effective are the Geonics EM series, which consist of a horizontally carried boom with a transmitter and receiver at either end, and electronics and controls between them. A varying magnetic field sent out by the transmitter generates electric currents in the ground which in turn generate magnetic fields detected by the receiver. A bonus of this type of instrument is that, operated at high frequencies of the order of 40 kHz (40 kilohertz = 40,000 oscillations per second),

Fig. 24a. Geonics electromagnetic instruments. The EM38 in Wadi Bahla, Oman.

they measure conductivity by looking at the component of the received signal out of phase with the transmitted signal (otherwise known as the quadrature component); but at low frequencies, typically 4 kHz, they look at the in-phase component which responds to magnetic susceptibility. Thus they can be two instruments in one. They will be discussed further in Chapter 4.

The two most suitable Geonics instruments are the EM31 and the EM38. The EM31 (Fig. 24b) is 4 m (13 ft) long, but has a claimed resolution of one-fifth of its length, 0.8 m (2.6 ft). In a few rapid experiments carried out in Britain, it does not appear to have been very effective in resolving features of such dimensions, but seems to be more so in drier conditions, as discussed below. The EM38 (Fig. 25), which is 1 m (3.3 ft) long, seems more appropriate to the scale of detailed archaeological investigations. It has the advantage that its maximum sensitivity when used normally is at 0.3–0.4 m (1 ft–1.3 ft) depth, so that

Fig. 24b. The Geonics EM31 at Woodhenge.

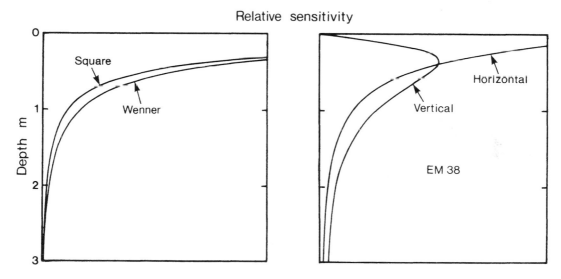

Relative sensitivity

Fig. 25. Sensitivity fall-off with depth for (left) two conventional resistivity configurations, compared with (right) the EM38.

surface interference is suppressed, and this sensitivity is only halved at 1 m (3.3 ft) depth. A detailed study comparing the EM38 with the normal RM4 0.5 m (1.6 ft) twin electrode resistivity system was made by Boucher (1987), who used two superimposed Roman forts at Elslack in Cumbria as a test-bed. The site is on boulder clay, and many of the defence works were constructed from sandstone of strongly contrasting resistivity;

there were also ditches showing more subtle negative anomalies. The results from the two surveys were closely comparable, and some anomalies showed more clearly in the EM38 survey. In general, high-resistivity anomalies showed more clearly with the RM4, and low-resistivity with the EM38. This is in keeping with the research into resistivity anomaly production described below, and the peak sensitivity of the EM38 being somewhat more than 0.3 m (1 ft) below the surface. Resistance measurement, which falls off in sensitivity from the surface, responds best to the superficial drying effect of high-resistivity features; conversely, drying tends to suppress the effect of low-resistivity anomalies near the surface, but not at the depth of maximum sensitivity of the EM38.

The comparative traverse at Woodhenge (Fig. 26) summarizes quite succinctly the relative virtues of the EM38 and conventional resistivity.

Electromagnetic instruments have also been shown to be highly effective, and superior to normal resistivity instruments, in the desert and semi-desert regions of the Middle East. There, the upper layers of soil can be almost completely dry for much of the year, and moisture is effectively confined to the deeper layers. Such conditions will defeat a normal resistivity meter with probes, but they positively favour electromagnetic instruments because the dry, sandy, upper layers do not impede the penetration of the signal to the lower layers. Alister Bartlett and I found the EM38 to be very effective for locating buried relatively damp deposits in Oman, and Frohlich and Lancaster (1986) report that the EM31 was successful in finding grave shafts about 1.5 m (5 ft) wide and 2 m (6.6 ft) deep at Bab ed-Dhra, Jordan. It could be that conductivity contrasts there are more extreme than can normally be expected in the generally wetter soils of temperate zones; and burial chambers leading off the shafts could have increased their detectability. The EM31, operated by Compagnie de Prospection Géophysique Française, was also successful in locating the bases of two small pyramids at Saqqara, at a depth of 8–10 m (26–33 ft) beneath disturbed rubble (Deletie *et al*, 1988). Here it was more successful than magnetic, VLF and DC resistivity. Detection at this great depth was by no means clear, and seems to have involved a careful consideration of the balance of probabilities, including the likely effect of the superincumbent disturbed material.

The last two techniques require further explanation. VLF makes use of the variable absorption of very low frequency (in this case, 16 kHz) signals from distant radio transmitters to produce a measure of apparent resistivity to depths of roughly 15–20 m (49–66 ft), from measurements of electric and magnetic fields over the site. It has shown satisfactory response to archaeological features (Tabbagh, 1973), although readings are biased by the direction of the transmitter, and its usefulness seems limited and rather academic in the archaeological context. DC resistivity is not subject to 'skin effects' which can reduce the effective penetration of AC resistivity measurements, and therefore it has rather better penetration. However, it utilizes non-polarizing electrodes that are relatively fragile and require careful emplacement, and it is therefore relatively slow and cumbersome to use.

Electromagnetic instruments clearly have much potential in archaeology. For resistance work in arid areas they are essential, as we have

Fig. 26. Comparison of measurements across the ditch of Woodhenge with a conventional resistivity meter and EM38. This was the standard test traverse used for the climatological study. 'Noisy' resistivity values (right) are probably due to chalk blocks in the remains of the bank causing poor contact. The EM38 responds smoothly and clearly to the bank, but strongly divergent readings are caused by metal rubbish near the surface.

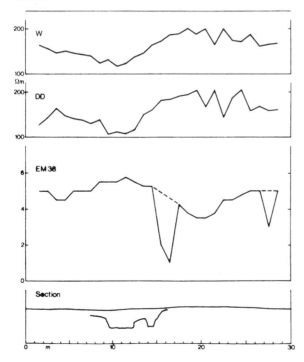

seen. At present, they are expensive compared with resistivity apparatus in temperate zones. They require careful setting up, but their response characteristics make them specially suitable for depth estimation. The EM38 is subject to drift and requires the addition of an improved read-out. It does have the potential to be used on the move for rapid recording, like the fluxgate gradiometer (Chapter 3), although it would need to be carried close to the ground to gain the benefit of its characteristic response with depth. It is also possible to use it for magnetic susceptibility measurements, although not very conveniently because readings at two heights have to be made at each point.

Resistivity and archaeology

If a high resistance feature lies buried in soil of uniform resistivity, much of the current will be forced to flow around it, finding longer and easier paths and upsetting the regular pattern of Fig. 18. This reduces current density in the vicinity of the feature, increasing the potential gradient. The gradient is sampled by the potential electrodes. Thus, in this case, V is increased so $V/I = R$ is increased, giving 'positive anomalies' in a line of readings.

A low resistance feature, such as a moist ditch filling, will provide an easy path which attracts the current. This lowers the potential gradient in the vicinity of the feature, giving 'negative anomalies'.

The first assumptions were that structures such as buried wall foundations would contain less moisture than the surrounding soil and therefore give high readings, or positive anomalies, while ditches, pits and comparable excavated features containing permeable soil silting, and cut into relatively impermeable or well drained natural, would give negative anomalies because of their high moisture content. This is very often so, but experience has revealed many complications. Resistivity detection depends on the interaction of the composition and geometry of features, electrode configurations and climatic variations, and the effects of all of these have had to be studied. As a result of quite considerable research effort, both theoretical and experimental, a technique that often seemed disappointing and unpredictable can now be used confidently and reliably, and sometimes with remarkable finesse.

Electrode configurations

The original standard configuration, inherited from 'big' geophysics and already introduced in this chapter, was the equally spaced current-potential-potential-current arrangement named after Frank Wenner who first published an analysis of its performance (Wenner, 1916). The Wenner arrangement seems obvious: two current electrodes to set up a potential gradient in the ground, and two potential electrodes between them to sample this gradient, just as the four-terminal measurement was made on samples in the laboratory. The formula for converting readings to apparent resistivity is $\rho = 2\pi aR$. The derivation of this, and the formulae for other configurations, can be quite simply worked out (Aspinall and Lynam, 1970).

The Wenner and related configurations are effective for many geological and geotechnical studies involving the horizontal layering of the ground, and the detection of large geological features and aquifers. But it was soon realized that they produced an unsatisfactory representation of some buried archaeological features, especially narrow ones, showing double or even treble peaks, or unduly broad ones. These effects are largely due to the fact that in soil the potential gradient between two current electrodes is steep in the neighbourhood of the probes, where the current is being crowded into a narrow cross-section of soil, but much less so in between (Fig. 18). Measurement is by no means tied to the Wenner configuration, and in fact an infinitude of different electrode arrangements is possible, some of which were shown to have advantages over the Wenner in some conditions. Perhaps the most comprehensive study of their relative merits to date was made by myself (Clark, 1980), and the following discussion is largely based on that work.

Much of this analysis depends on the simulation of archaeological surveys in an electrolytic tank, supplemented for illustration by two-dimensional field plots obtained with conductive paper. In the tank, which was less than 1 m (3.3 ft) long, water made conductive with salt took the place of the soil and simulated archaeological features were moved under fixed electrodes arranged in appropriate patterns. This reversal of the real-life situation avoids the 'end-effects', due to the limited volume of the tank, that would occur if the electrodes were moved. A

constant current meter with continuous automatic plotting was used, so that the traces were an
exact representation of the instrument response.
Many variations were tested and only a selection
of them, regarded as the most significant, is
shown and discussed here (Figs. 28–31). They
were all made over simulated linear features,
representing mainly walls and ditches, and show
two peaks. Normally the simpler of each pair of
peaks is with the electrodes lined up parallel to
the feature, while the 'end-on' response tends to
be the more complex and interesting, and is the
one normally discussed below. The peak sizes
shown indicate a greater sensitivity to low resis-

Fig. 27. Electrode configurations. W = Wenner; DD =
double dipole; Tw = twin electrode; Sq = square
array; Sch = Schlumberger; P = Palmer; a = standard
equal spacing. Those found best for archaeological
detection are above the line. Schlumberger and Palmer
are shown as arranged for the tank tests, but the paired
electrodes are often used closer together.

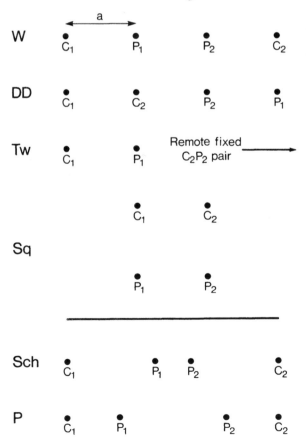

tivity than to high resistivity features, although
the simulated features were either almost perfect
insulators or conductors – ideals not likely to be
encountered in the ground. Isolated 'discrete'
features, such as pits, were also tested but they
tended just to give reduced versions of the linear
responses. The configurations discussed are
shown in Fig. 27.

The Wenner peak over a deep-sectioned feature (Figs. 28, 29) is broadened by a strong response between each current-potential pair as the
steep part of the gradient associated with them
(Fig. 28) passes over the feature. With a low resistance feature, these tend to resolve into separate
peaks, producing a shouldered effect (Fig. 31).
This can give a false impression of feature width.

More troublesome for survey interpretation is
the simple double peak often encountered with
high resistance features (Fig. 31). This sometimes
appears in grotesquely extreme form, as in a survey in Spain by Linington (1967). The electrolytic
tank showed that this effect occurs when the
archaeological feature behaves as a flat horizontal sheet, or lamina. This was surprising, because
it is seen over many features that clearly are not
laminar; in combination with climate studies, it
has led to an important insight, described below,
into how a large proportion of positive archaeological resistivity anomalies are generated. A
high resistance lamina gives a double peak with
the Wenner arrangement because it upsets the
potential gradient pattern between each pair of
current and potential probes, while hardly disturbing the almost horizontal current lines that
create the relatively slight potential gradient between the potential probes. This is illustrated by
the conductive paper profiles in Fig. 32A and B,
which should be compared with the undisturbed
potential pattern in Fig. 18.

In contrast, tank experiments with a low resistivity lamina showed that it gives much the same
response pattern as a solid body, with a major
central peak and two side shoulders. This must be
because the easy path it provides is as strong an
attraction for the current as the top of a solid
body. The current pattern is distorted so that the
potential gradient between the potential electrodes is strongly affected (Fig. 32C). Thus the
anomalies caused by high and low resistivity features are not necessarily mirror images of one
another.

Three electrode configurations have been used

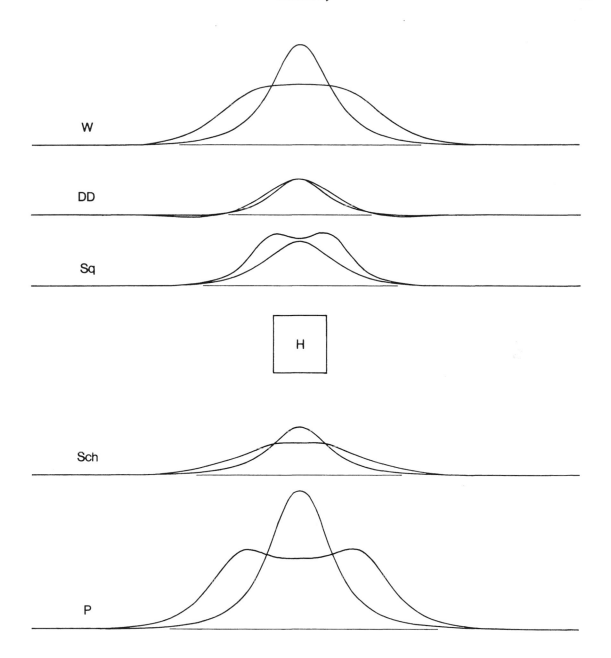

Fig. 28. Electrolytic tank simulation of response patterns for different electrode configurations over a square section high-resistivity feature (a wall foundation, for instance) of the same width as the electrode spacing and depth equal to half this. Single peaks in each case are with current flow parallel to feature length. The configurations shown below the feature diagram are those not recommended.

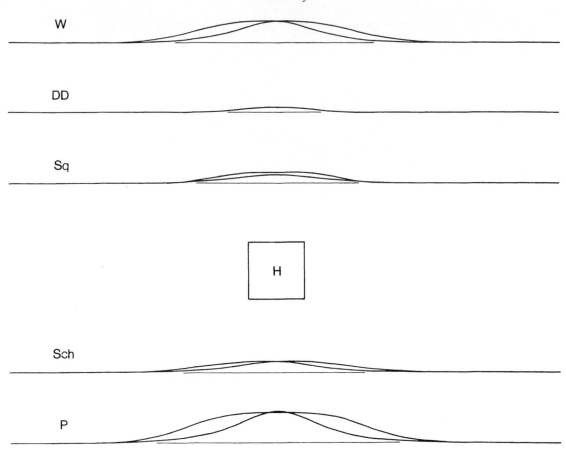

Fig. 29. As Fig. 28, but with the feature at double depth (= standard electrode spacing).

successfully to combat the widening and multiple peaking problems. They share in common the exploitation of the steep potential gradient and consequent enhanced sensitivity between each current–potential pair. Their performance in relation to the Wenner is shown in Figs. 28–31. Two less successful attempts to solve the problems, Schlumberger and Palmer, are shown at the bottom of the figures.

Fig. 30. (Opposite) Tank response patterns for a laminar high-resistivity feature at a depth of half the electrode spacing (top) and electrode spacing depth (bottom).

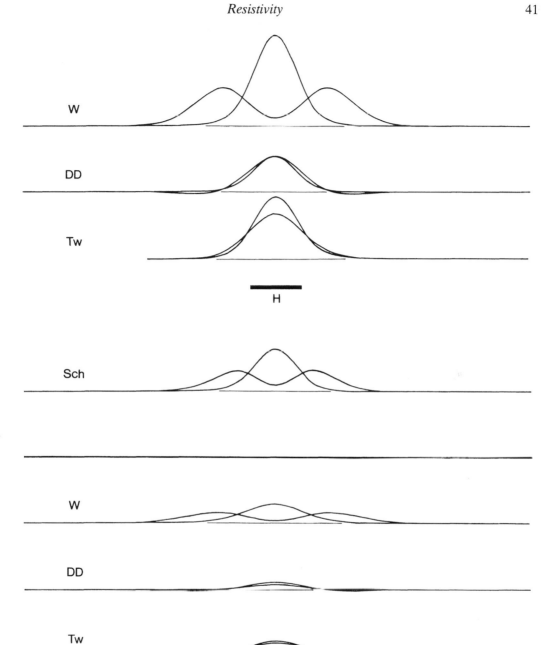

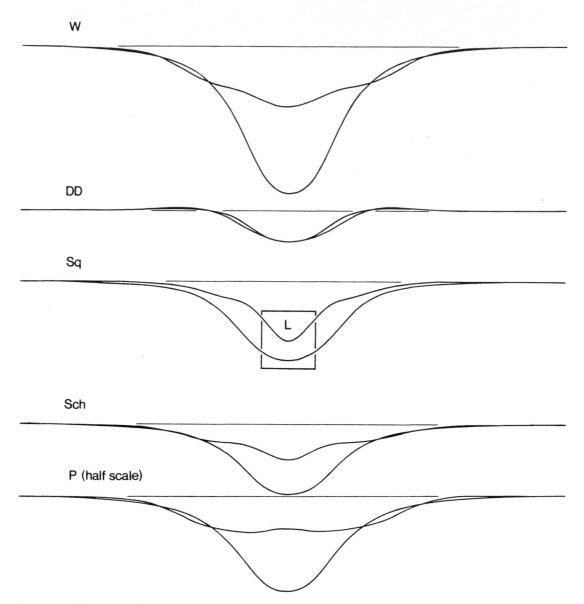

Fig. 31. As Fig. 28, but for a low resistivity feature.

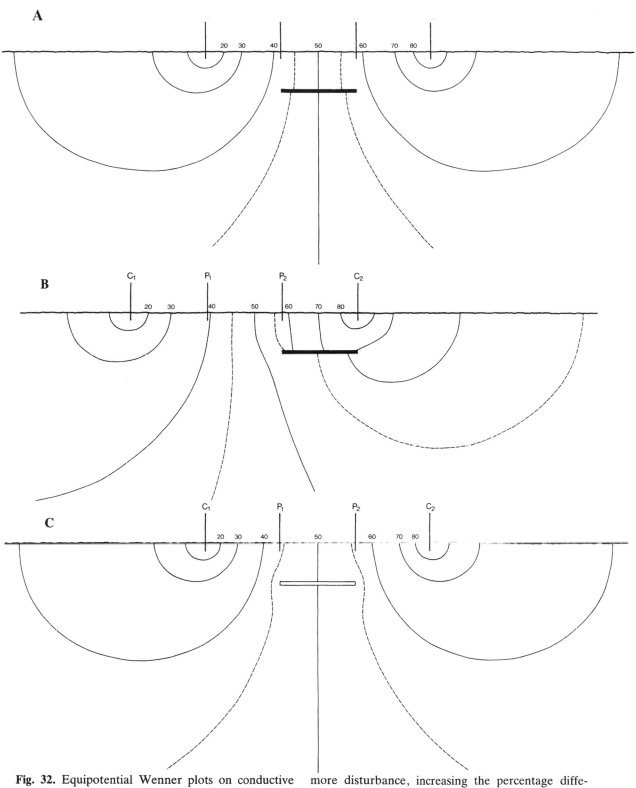

Fig. 32. Equipotential Wenner plots on conductive paper. Compare Fig. 18 and the top diagrams of Figs. 30 and 31. (A) A high resistivity laminar feature at the centre of the array causes little disturbance of the current pattern. (B) Beneath P_2 and C_2 it causes much more disturbance, increasing the percentage difference in potential between P_1 and P_2. (C) The central conducting lamina is much more disturbing than the non-conductive.

Double dipole configuration This is sometimes known as the Wenner β, and is achieved by simply interchanging one current and one potential connection on the standard Wenner (Wenner α) which makes possible the W/DD change-over switch on the Martin-Clark meter (Fig. 33). The

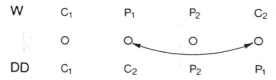

Fig. 33. Simple switching required to interchange instrument connections between Wenner and double dipole.

result is a configuration that is most sensitive between C_2 and P_2, because P_2 is sampling the steep gradient close to C_2, while the sensitivity between C_1 and P_1 is very small because of the remoteness of P_1. Thus the sensitivity of this configuration is concentrated between C_2 and P_2, and its responses are normally clear and simple. A drawback is that, because the current probes are close together, current penetration is limited and its sensitivity and depth response are only about one-third that of the normal Wenner. Resistivity in a uniform earth, $\rho = 6\pi aR$. Benefits of a Wenner/double dipole change-over switch are clear and unambiguous definition of relatively shallow features with double dipole, and better response to deep features with Wenner – with the added advantage that the Wenner response tends to become less complex with depth. We shall see

Fig. 34. Use of the Wenner/double dipole switch in the field. Traverse across two Roman buildings on clay (note low background readings) near St Albans. The definition of the walls by double dipole avoids the confusion of Wenner.

that double dipole can be used for very subtle work on shallow features. A good example of the two configurations used in combination is shown in Fig. 34.

Twin electrode configuration Developed specially for archaeology, this configuration has the distinction of being so specialized that, until recently, no other application seems to have been found for it. It is, in effect, the Wenner configuration divided into two, with a very large separation between the CP pairs, which removes the double peaking problem discussed above. One of the pairs is placed in a fixed position, while the other two are moved over the site and act as the detector probes. Simple mathematics (Aspinall and Lynam, 1970) show that, if the separation of the pairs is at least 30 times their individual spacing, then variations in the separation will affect readings by less than 3 per cent, which is insignificant in resistivity work, especially if the data are filtered for presentation. Once they are separated this much, the relative orientation of the probe pairs is also irrelevant and penetration is improved, because the current flow from the current probes is effectively radial. To achieve the same penetration as, for instance, the normal Wenner, the probe spacing need be only about half as great. Thus a 0.5 m (1.64 ft) spacing with twin electrode 'sees' as deeply as the normal 1 m Wenner, but with greater horizontal resolution. Such a compact array lends itself to construction as a frame with the instrument (and optionally a data logger) attached, and is the recommended standard way of using the Geoscan RM4 (Fig. 35). This configuration is only shown in one of the tank diagrams (Fig. 30), because most of the research had been completed by the time it was developed, but its clarity of response relative to

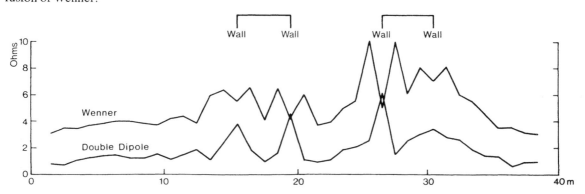

Fig. 35. Survey at Avebury. The twin electrode system as used by the Ancient Monuments Laboratory, with Geoscan RM4 resistance meter and data logger. The logger is triggered automatically to record a reading when the probes make contact with the ground. (Top, left) Two-handed operation; readings are taken 0.5 m (1.6 ft) from guide line. (Top, right) Single-handed operation, a quick method for good contact conditions. In the background, a second line is being moved to the next position. (Bottom) Dumping the readings into a computer for storage and display on the screen as traces.

Wenner, and signal strength relative to double dipole, are clearly shown.

Disadvantages of the twin electrode can easily be imagined. The separation of the current electrodes is so great that it should respond to geology rather than near-surface phenomena. It is indeed affected by geology, which contributes to the rather high 'background level' of readings. However, the response to it is normally insignificantly variable compared with any archaeological effect because the geology lies on the flat and insensitive part of the potential gradient while archaeology affects the steep part of the gradient close to the current probes, which we have already seen to be very sensitive.

Calculating resistivity values with this configuration is difficult, because the fixed as well as the moving probes contribute, so that only relative resistance readings are normally used. In fact, with the dual spacing probe arrangement mentioned below and discussed in Chapter 7, it is possible to estimate the apparent resistivity beneath the moving electrodes from the following formula, derived from Appendix I in Clark (1980):

$$\rho = \frac{2\pi\ ab(R_1 - R_2)}{b - a}\Omega\text{-m}$$

where a and b are the narrow and wider probe spacings, and R_1 and R_2 are the larger and smaller resistance readings. If the two spacings are 0.5 m and 1 m (1.6 and 3.3 ft), this simplifies to

$$\rho = 2\pi(R_1 - R_2)\Omega\text{-m}.$$

The assumption is made that the apparent resistivity is the same for both probe spacings.

Convenience, speed in use and practical effectiveness outweigh most of the objections to the twin electrode system, and it has become deservedly the most popular resistivity method in archaeology. It can be home-made, as suggested above, or sophisticated (Fig. 35). A practical constraint is that a survey is tethered to its fixed electrodes, so that it is most useful for area surveys. Long linear 'search' traverses are more conveniently conducted with Wenner, preferably combined with double dipole, using the leapfrog method. The twin fixed electrodes can certainly be relocated to cover new ground, but this necessitates adjusting their spacing to keep the background level of the readings the same as for those already measured (Fig. 36). The time and possible errors involved in this are justified for area surveys, but not for single line traverses.

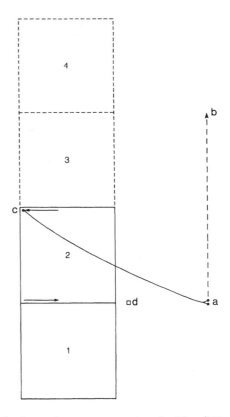

Fig. 36. Twin electrode survey procedure for 30 m (100 ft) squares. With the instrument mounted on the electrode frame, as with the RM4, and about 75 m (250 ft) of cable, squares 1 and 2 can be surveyed with the fixed probes at a. The moving probes are left at the last reading point, c, and the reading noted. The fixed probes are then transferred to b, for squares 3 and 4, and their separation is adjusted to bring the reading at c to its original value. Another method of surveying, often used with the Martin-Clark, is to keep the instrument at d, with separate 50 m (160 ft) cables to the two sets of electrodes.

The square array This was first constructed as an empirical solution to the complex response to small features obtained with the Wenner array, on the assumption that a more compact configuration of electrodes should give a sharper response. The concept appeared in a paper on the measurement of the resistivity of semiconductors on the microscopic scale (Uhlir, 1955), and was first tested for archaeology by myself in the mid-1960s. At the same time Habberjam was independently developing it for use on the geological scale. The array could be made up as a table supporting the instrument, and switching

Fig. 37. The original square array with 0.76 m (2.5 ft) spacing. The electrode legs are small steel angle, squeezed flat at the ends to be vertical and easy to insert, with the added bonus that the blade-shaped profile has a larger area/length ratio than round rod. The meter stands on top, and the switch on the side interchanges the electrodes to permit averaging. Assembly with wing nuts makes it fully demountable.

Fig. 38. Using the square array to locate the Hog's Back barrow ditch.

made it possible to interchange electrode functions so that readings at one point could be averaged to remove any directional bias that might exist. Theory showed the depth sensitivity to be not much less than that of Wenner at the same spacing (Fig. 25).

Some very successful surveys were carried out with a square array of 0.76 m (2.5 ft.) spacing (Fig. 37), a lightweight structure almost as convenient to use as the twin electrode (which had not, in fact, been developed at that stage). The electrolytic tank experiments confirmed the effectiveness of the array, and also confirmed the necessity for the averaging of at least two readings at right angles to remove twin peaking, for instance when the PC pairs were parallel to a narrow linear feature (Fig. 28). It is perfectly easy to take four readings at each point, but there is a Principle of Reciprocity which states that the current and potential electrodes in any resistance measurement can be interchanged without changing the reading, so that only two readings are necessary in good contact conditions.

The square array has latterly been overshadowed by the twin electrode, with its deeper penetration per unit probe spacing, ability to produce unambiguous readings without averaging, lighter weight and smaller number of contacts. Even so, the square array is the most compact contacting resistivity unit so far devised, with the freedom of movement of the leapfrog system but without its complications or trailing leads. It has been revived by the Centre de Recherches Géophysiques for use with their tractor-mounted resistivity systems (Hesse *et al*, 1986). Apparent resistivity $= (2/2 - \sqrt{2})\pi aR$.

Other electrode configurations A number of electrode systems have been advocated over the years. In particular, two variations on the Wenner theme have been applied in archaeology, and are included in Figs. 28–31. In the *Schlumberger*, the two potential electrodes are moved closer together, which it was originally thought would improve the definition of small features. The tank experiments show that this is true for relatively shallow features of solid section, but in other cases it merely reduces the size of the reading without improving definition because the closer electrodes measure a smaller part of the same potential gradient.

The *Palmer* configuration (Palmer, 1960)

attempted to tackle the problem of signal size rather than spatial definition by using an approach exactly opposite to that of the Schlumberger. The potential electrode spacing was widened instead of narrowed so that both were in the steep part of the gradient close to the current electrodes. To facilitate its use, Palmer joined the CP pairs with rigid insulated links which served to maintain the spacing and also as handles. Working along a traverse in the normal 'end-on' sense, the tendency to see each feature twice was greater than with Wenner. To overcome this, he advocated the 'broadside-on' approach (note the simple peak this produces on the tank traces), but this was of limited value because it was really only effective for linear features and required a prior knowledge of the direction in which these were running! But Palmer was close to the concept of the twin electrode configuration.

Climatic effects and the nature of resistivity anomalies

As resistivity detection depends upon the distribution in the ground of water precipitated from the sky, it was realized early on that response must be conditioned not only by the nature of the remains and the soil containing them, but also by the 'water balance' – precipitation input and water loss by 'evapotranspiration', a combination of evaporation and take-up by vegetation. Experiments to quantify these effects require an undisturbed site containing features of known composition and size, available for at least a year, and precipitation data from a local weather station. Measurements are repeated along the same fixed traverse line at intervals of usually a month.

The first such experiment was made by Al Chalabi and Rees (1962) at Wall in Staffordshire,

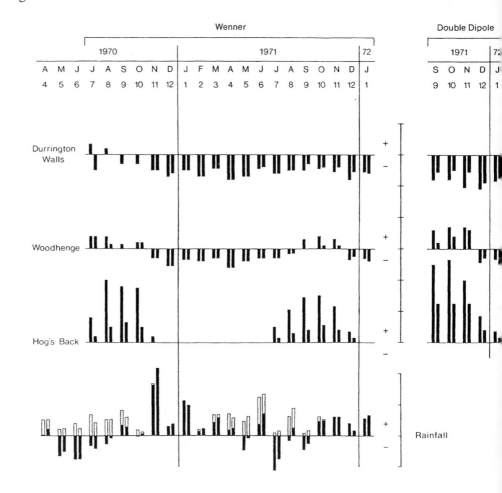

over a series of Roman defensive ditches cut into soft Triassic sandstone. It was found that the ditches remained detectable throughout the year as uncomplicated low (negative) resistivity anomalies, which even at their worst were about half as large as at their best. The clearest response was after a period of hot, dry weather when the soil had suffered an overall loss of water by evapotranspiration. Comparison with weather statistics showed that, as a rough rule, best results would be obtained in July, but that good conditions would be very likely to prevail from June to September.

The chalkland experiments Many archaeological sites lie upon chalk bedrock, yet its resistivity behaviour often seemed anomalous. It therefore seemed important to understand why, and I ran a chalkland experiment from 1970 to 1972 (Clark, 1980). Fixed traverse lines were laid out across the great ditch of the henge monument of Durrington Walls and the intermediate-sized ditch of the adjacent Woodhenge, both in Wiltshire, and across the relatively small ditch of a bell barrow

on the Hog's Back ridge in Surrey – the specific formation in each case being Upper Chalk. Control was provided by nearby archaeological sections, the Wiltshire sites by Wainwright and Evans (Wainwright and Longworth, 1971), and the Surrey site by myself. All were filled predominantly with ploughsoil and ploughwash, with coarse chalk primary silting beneath. It became apparent that the response to these ditches varied greatly according to their size and the time of year. The results are summarized in Fig. 39.

These variations can be broadly ascribed to the relatively coarse texture of the ditch fillings compared with the closer texture of the natural chalk into which they were cut. When small in cross-section, the fillings will gain and lose water quite readily, whereas the chalk, although relatively permeable in a geological sense, tends to be more stably retentive than the fillings. In contrast, the very large ditch of Durrington Walls almost always gives a negative anomaly, showing that the massive cross-section of its filling is acting as a very effective water reservoir. Only after the driest conditions, in July and August 1970, did the anomaly go positive, and then only with the narrower 1 m (3.3 ft) electrode spacing with its

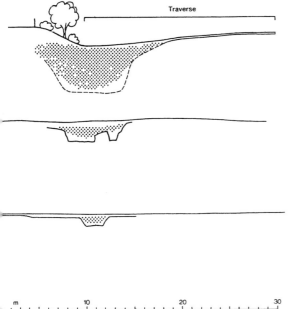

Fig. 39. Summary of resistivity tests on ditches in Upper Chalk bedrock, + high, − low anomaly relative to background level. First bar of each pair, 1 m (3.3 ft) electrode spacing; second bar, 1.5 m (5 ft) spacing. Vertical scale interval 100 ohm-metres. Water balance is shown at the bottom. Open parts of bars show rainfall; black is net gain or loss of water after allowing for evapotranspiration. First bar of each pair, Durrington Walls and Woodhenge; second bar, Hog's Back. Vertical scale interval 100 mm (4 in.).

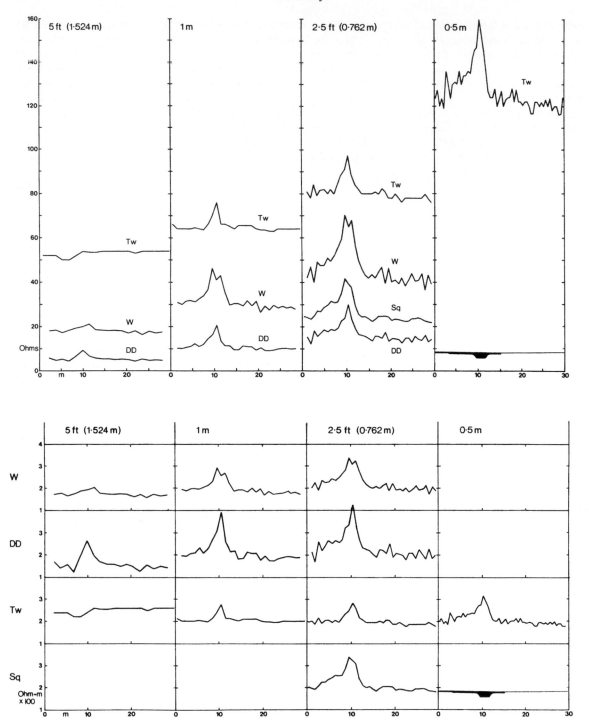

Fig. 40. (Top) The Hog's Back barrow. Measurements on 13 November 1971 repeated with various configurations and probe spacings, compared with the profile of the ditch (bottom right). Raw data. The 'noisiness' of twin electrode is probably due to using separate probes. (Bottom) The same, reduced to resistivity values. By 17 January these anomalies had disappeared.

shallower penetration, revealing net water loss only from the top of the filling.

At the other end of the scale is the ditch of the bell barrow on the Hog's Back. At 2 sq. m (22 sq. ft), its cross-section is only about 4 per cent of that of the Durrington Walls ditch, and of a size more typical of the prehistoric ditches most frequently found. This small size, and its uncomplicated form, have made possible a broadly applicable analysis of the behaviour of small resistivity anomalies.

The picture here was radically different from that of the big ditch. During periods of net water gain, saturation or field capacity is reached; the mean resistivity becomes indistinguishable from that of the chalk into which the ditch was cut and it is undetectable, or very nearly so. But when conditions of water deficit develop in the drier months of the year, its resistivity anomaly rapidly increases, becoming much greater than that of the big ditch at the narrower 1 m (3.3 ft) electrode spacing. The fact that its anomaly is positive suggests that the ditch filling has dried out, but a series of measurements taken on one day (Fig. 40) tells a more detailed story. At 1.5 m (5 ft) there is little response, indicating that the whole cross-section of the ditch may not be contributing to the anomaly. At 1 m (3.3 ft) and 0.76 m (2.5 ft) the anomaly is strong and increasing in size, showing that the response is to something relatively shallow. At both these spacings, the

Fig. 41. The Hog's Back ditch section (top), and approximate form of vertical water content/resistivity contour distribution (bottom) deduced from Fig. 40 and the tank tests on laminar features.

response is double-peaked, so its cause must be laminar -- and apparently decreasing in width to match the electrode spacing. A moisture regime that would explain all these indications together is shown in Fig. 41. It would seem that the anomaly is due not so much to the ditch filling itself as to the fact that its relatively coarse texture blocks the capillary movement of moisture from the natural chalk to the topsoil, while at the same time this moisture maintains the lower part of the filling at a fairly low resistivity. The extreme sensitivity of resistivity in these conditions is shown by the response to the berm by the Wenner, double dipole and square arrays at 0.8 m (2.5 ft), although the soil covering it is only 10–14 cm (4–6 in.) deeper than the soil on the other side of the ditch. That a twin electrode responds only at 0.5 m (1.6 ft) spacing confirms its equivalence to wider spacings with the other arrays.

Another example of the extreme sensitivity of resistivity to topsoil depth in conditions of water deficit was observed recently close to the Roman villa at Compton, Surrey. Clear linear positive anomalies were interpreted as building walls, but on excavation proved to be shallow gullies defining the upper edges of cultivation terraces. This survey will be analysed in Chapter 7.

The response pattern of the intermediate sized ditch of Woodhenge is a rather alarming combination of the behaviour of the large and small ditches, the anomaly sometimes being predominantly positive and sometimes negative. The deeper parts of the ditch tend to follow, in diminished form, the behaviour of the great Durrington ditch, whereas in the shallow parts over the

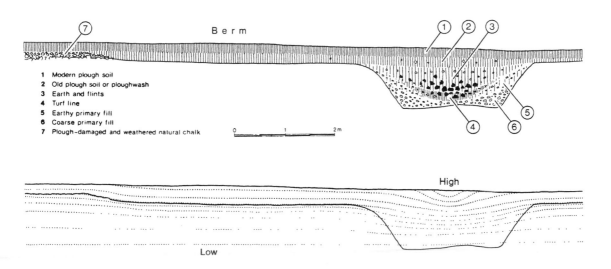

1 Modern plough soil
2 Old plough soil or ploughwash
3 Earth and flints
4 Turf line
5 Earthy primary fill
6 Coarse primary fill
7 Plough-damaged and weathered natural chalk

undisturbed natural chalk 'island' and the gentle inner slope high anomalies appear in the dry season, in the manner of the small ditch. Although epitomizing the problems of resistivity prospecting, on chalk at least, such behaviour can yield valuable information to the subtle practitioner, as briefly discussed below.

Resistivity and crop marks The crop mark behaviour of the larger ditches on chalk, and probably most lithologies, parallels their resistivity response quite closely. Smaller ditches, however, seem able to continue to support a lush, positive mark at the same time that resistivity anomalies are beginning to rise in conditions of water deficit. This must mean that the better rooting conditions provided by the fillings, combined with deeply stored water, are the dominant effects on the crop, while resistivity measurement readily penetrates the relatively damp chalk, detecting the water in it which is inaccessible to the roots except as it gradually reaches the surface of the chalk mass. Thus the upper filling of the ditch, depleted of moisture by the roots themselves, stands out in contrast as a high resistivity anomaly. The moisture reserves of the larger ditches, however, tend to counteract this imbalance, most effectively in the case of the massive ditch of Durrington Walls.

That the crop roots do gain some water diffusing from the chalk is supported by observations at Woodhenge and many other sites. On the first air photographs taken of the site by its discoverer, Squadron Leader G. S. M. Insall, VC, on 30 June 1926, both sides of the crop mark of the ditch are darker than the centre (Fig. 42). This is not in keeping with the resistivity experiment, in which the response lacked this symmetry. Also on the photographs, the comparatively insubstantial post holes and slighter ring ditches (probably of barrows) with fillings as a whole in closer contact with the natural chalk, are all dark. In extended drought these conditions break down and all ditches and pits become parched. This has occurred in a photograph of the Woodhenge field taken by P. Goodhugh at the height of the drought of 1976 (Fig. 43). All the circles and the ditch of Woodhenge appear as parch marks of

Fig. 42. (Above) Woodhenge crop marks, 30 June 1926. (*Crown copyright*)
Fig. 43. (Below) Woodhenge crop marks, summer 1976.

dramatic clarity, although the Woodhenge ditch still retains signs of a thin dark rim. This photograph also reveals many slighter ditches, probably of Romano-British fields, never previously seen. Thus it seems that a ditch may be too small to produce a positive growth mark, but still capable of causing a parch mark in conditions of extreme stress – which is reminiscent of the resistivity behaviour of the Hog's Back ditch. The fact that these marks existed at all demonstrates the ability of the undisturbed chalk to retain and deliver moisture steadily to roots in the rest of the field, probably largely in vapour form.

Other climate experiments Less exhaustive experiments have also been conducted over other lithologies. A study was made on *London Clay* of the levelled remains of a farm under grass on the campus of Surrey University during 1975 when the summer was exceptionally dry. Natural clay is very water-retentive and gives low background readings against which structures such as building foundations can be expected to show in sharp contrast. This was the case here, the background level never exceeding 15 Ω-m, in spite of the drought. The strength of response followed a pattern very similar to that of the small Hog's Back ditch on chalk. The best contrast was from June to November, peaking in September when anomalies reached 214 Ω-m above background with 1 m (3.3 ft) Wenner, and even higher with double dipole (296 Ω-m), hinting that the remains were shallow in depth. It was at its weakest in May (W = 12 Ω-m; DD = 18 Ω-m), just before the commencement of the rise in mid-June. In February, March and April, it was at a quite constant level of about W = 15 Ω-m; DD = 23 Ω-m. Although the anomaly never disappeared, as did that on the Hog's Back, it was only in the driest conditions on 7 September that some slight additional anomalies appeared, possibly caused by narrow wall foundations. By 15 September, after heavy rain for several days, these again disappeared and the main anomaly was reduced from the peak level given above to W = 82 Ω-m; DD = 113 Ω-m.

An experiment was carried out in central France by Hesse (1966), on a *limestone* bedrock with a soil cover about 30 cm (1 ft) deep. A test traverse, for which 1 m (3.28 ft) Wenner was used, was laid out across a feature which subsequently proved to be a covered, air-filled stone sarcophagus in a grave pit. Its anomaly was always high with respect to the background, and at its best in the form of very distinct twin peaks. Maximum response again occurred during the latter part of the year, especially from August to October.

Ice has a very high resistivity and frozen soil can make measurements difficult or impossible. Hesse's experiment coincided with one of the coldest winters of the century. After ten days of heavy frost from 11 January 1963, when the ground temperature remained between 0°C and -10°C, measurements showed a rise in base level from about 80 Ω-m to 500 Ω-m, and the anomaly due to the feature was largely obliterated by erratic peaks, termed 'noise'. The freeze continued almost uninterrupted until 5 February, and even on 12 February the resistivity base level had only fallen back to about 250 Ω-m and the noise effects were still stronger than the anomaly caused by the sarcophagus. On the other hand, the effect of a more normal period of frost in December 1962, lasting four days with minimum air temperatures between –5°C and –10°C, was not detectable. It is likely that the difficulty was due to the frost causing high contact resistance rather than to a more profound effect on the feature itself. It was probably exacerbated by the continental climate of central France; Russell (1957) observes that in the fairly typical English conditions at Rothamstead, Herts, there is very

little temperature variation between summer and winter at depths as small as 20–30 cm (8–12 in.), and that a grass cover has a considerable effect in smoothing out extremes of temperature.

Summary of optimum resistivity conditions The periods for best detection of features of various sizes and types on different bedrocks are summarized in Table 1. It shows that, with the exception of the larger ditches on chalk, there is a quite close similarity in periods of optimal response for a wide variety of lithologies and patterns of weather, with a 'core time' common to all the sites of July–September. However, apart from the chalk, conditions are not normally critical because features (except the slightest, or when heavy and prolonged freezing produces 'permafrost' conditions) are detectable throughout the year. This may prove to be broadly applicable to north-west Europe, and possibly the temperate zones generally. An important reservation is that the complications encountered with chalk could have emerged because of the thoroughness with which it has been studied, and may not be unique.

Further experiments are required to test such possibilities, and the more neglected lithologies, for instance gravel, although in the meantime it may be valid to assume that its behaviour would be similar to that of the sandstone at Wall. The

Table 1. Optimum times for detecting various feature types on different lithologies.

Site	Bedrock	Feature	Dimensions	Anomaly Type	Mean Best Months	Peak Months
WALL Staffs	Triassic Sandstone	Ditches	Width variab. Depth 3.4 m max.	Low	June – September	July
HOG'S BACK Surrey	Upper Chalk	Ditch	Width 2.5 m Depth 1.1 m	High	July – November	September
SURREY UNIVERSITY	London Clay	?Rubble ?Walls	Width c.10 m & 0.5 m Depth ?	High	June – November	September
POUILLY Nievre	Limestone	Stone coffin	Width c.0.5 m Depth c.1.5 m	High	?July – October	?October
DURRINGTON WALLS Wilts	Upper Chalk	Ditch	Width 17.7 m Depth 6.0 m	Low	December – June	March – April
WOODHENGE Wilts	Upper Chalk	Ditch	Width 6.3 m Depth 2.1 m	Low/High	December – June	March – April

types of feature tested have been restricted, with a heavy emphasis on ditches. Work has not been done on pits, but they may probably be considered similar to ditches, providing allowance is made, for instance, for the better water retention of those with closer-textured fillings often associated with intensive occupation. There has also been remarkably little long-term study of the response to building foundations, mainly, it would seem, because of a lack of suitable and convenient test sites. Much depends on the relative porosity of the building material and the ground containing it. With flint walls, on chalk at least, conditions for this type of feature are not highly critical. The Roman building at Boscombe (see Chapter 4) was surveyed in continuous rain in February, yet gave anomalies up to $120 \, \Omega$-m, and building foundations on a clay site awash with rainwater were easily detected. Brief experience on sand, however, indicates that stone foundations may become increasingly difficult to detect as the sand dries out and approaches the high resistivity of the stone, so that wetter conditions may be more suitable in this case, as for the large ditch on chalk. The converse of this was observed in Painshill Park, Surrey, on the site of the remarkable eighteenth-century landscape feature known as the 'Turkish Tent'. All that remained of this was a brick floor set in the natural sand, and the floor gave a negative anomaly in dry conditions because the relatively close-textured brick was retaining moisture better than the sand. I was prepared for this interpretetion by a debacle in a lecture a short time before. I had hoped to demonstrate the high resistivity of a brick by passing it under electrodes in water. The demonstration had worked very well when rehearsed, but nothing happened when it was re-run for the lecture: the brick had been left in the water and become saturated. I try to console myself that this was a more telling demonstration than a successful one, and it certainly helped with Charles Hamilton's Turkish Tent.

There is an infinity of subtle variations from region to region, and site to site. At Cottam, on the chalk of the Yorkshire Wolds, barrow ditches similar to that on the Hog's Back, but rather deeper, gave clear negative anomalies, which were never obtained on the Hog's Back. This was probably due to a combination of differences. The soil in Yorkshire contains a boulder clay element which gives it a closer and more water-retentive texture than that on the chalk in the south. Then there is the hardness of the chalk, caused by filling of the Foraminifera cells which make up half its bulk with calcite (Sorby, 1879), whereas in the southern chalk they are empty. This must reduce the relative water capacity of the Wolds chalk, as well as generating less primary silt and allowing a greater proportion of topsoil in the ditch fillings. In spite of these variations, it is likely that these ditches follow the general pattern of increasing anomaly with increasing water deficit. The cross-section of a ditch is probably more crucial than the properties of its constituents.

It is difficult to choose the ideal time to survey a site on chalk which has ditches and other features of different or unknown size. One survey during the first half of the year and another in the second would be ideal, but is often impracticable. For a single survey one must choose the second half of the year when all sizes are detectable even though the large are relatively weak and the intermediate unpredictable. Providing there is not too relentless an onset of rain, as there was during the field experiment in November 1970, good results may be obtained in December when the larger ditches are showing reliably low anomalies and those of the small ditches still remain high. At Woodhenge, the high anomalies caused by the unexcavated chalk 'island' and a last remnant of the bank revealed the presence of the ditch when the deeper filling was least detectable in summer and autumn. From December 1970 to May 1971 the island was lost in the normal low ditch anomaly so that surveys at different times of the year over such a ditch could be used to ascertain its outline as well as information about its more detailed structure. In this way, as with the use of configurations or probe spacings to define the berm of the Hog's Back barrow, resistivity can be used to obtain information of great subtlety.

The production of anomalies can be predicted most reliably if the soil water balance for the area is monitored. Information about this can be obtained from some weather stations or in Britain from the Meteorological Office in Bracknell. As already discussed, high-resistivity anomalies on chalk (small ditches) seem to increase as long as the water deficit conditions favourable to their production persist, while low resistivity anomalies (large ditches) tend to level off after about four months of water gain when saturation or

Feature & Conditions	Probe Spacing	Time from Onset	Remarks
Large ditch (DW) (Water gain)	1 m 1.5 m	1 month 2 months	Not critical except after very dry periods
Medium ditch (WH) (Water gain)	1 m 1.5 m	3 months 3 months	Long delay because of need to wait until high anomaly is reversed
Small ditch (HB) (Water deficit)	1 m 1.5 m	1 month 2.5 months	Long delay at 1.5 m because this spacing is unsuitably large for the feature

Table 2. Times from onset of favourable conditions to appearance of clear anomalies, for two Wenner electrode spacings.

'field capacity' is reached. For practical purposes, it is probably best to consider the time lapse required from the onset of favourable conditions to the production of a clear anomaly. If this anomaly is (arbitrarily) defined as 25 Ω-m with Wenner, the times for 1 m and 1.5 m (3.3 ft and 5 ft) probe spacings are as given in Table 2. This will also apply to probe spacings of equivalent penetration with other configurations, with some allowance for the fact that, for instance, double dipole gives a more effective definition of small features than does Wenner.

Thus it can be seen that the development of resistivity anomalies depends on precipitation input or deficit over a matter of months. Rain during the week or so before (or even during) a survey does not affect the basic moisture distribution producing the anomaly, although its superficial short-circuiting effect will alter the background resistivity level and may temporarily suppress the apparent size of anomalies.

The whole of the discussion above refers to soil with vegetation cover, mainly grass. Bare soil lacks the transpiration factor and so would be likely to inhibit somewhat the development of positive anomalies in dry conditions. In fact, bare soil is rare except on ploughland in winter, the part of the year when these conditions do not normally occur. It is perhaps most noticeable when making resistivity surveys in gardens and parks where readings obtained over flower beds will differ from those over turf, and allowance must be made for this. Even a lightly used track across a grass field can have a very noticeable effect: compression of the soil by traffic tends to reduce its surface area and damage to vegetation to reduce evapotranspiration, which combine to cause lower readings. Readings can become high in the vicinity of trees, which draw a great deal of

water from the ground as well as reducing input with their umbrella-like canopies. Background levels of resistivity surveys are also much more affected by changes of geology and of soil depth than those of magnetic surveys, which are largely 'self-filtering'. Ways of coping with these problems are discussed in Chapter 7.

Finally, perhaps the clearest lesson of all these soil moisture studies is that resistivity contrast depends upon contrast in porosity. For example, one may be disappointed in seeking ditches in clay or alluvium which may well be silted with the kind of material into which they were dug, or sandstone building foundations on a sandy subsoil. On the other hand, the clay backfilling, especially, may not regain its original compaction and remain chunky, so may still be detectable because of the insulating air gaps. One must attempt to judge such possibilities for each individual site.

Some resistivity practicalities

Survey layout and procedure The general principles are described in Chapter 8. When using a linear four-probe configuration, it is worth bearing in mind that the readings are assumed to apply to the mid-point of the array of four active probes. This means that, for instance with 1 m (3.3 ft) spacing, the first reading is at 1.5 m (5 ft), and the last is 1.5 m short of the last probe position. Therefore, when using this type of survey with the standard grids described in Chapter 8, the traverse lengths must be extended beyond the grid lines by one reading spacing at each end to achieve complete coverage.

Twin electrode procedure is shown in Figs. 35 and 36, and more will be said about it in Chapters 7 and 8. It is worth adding here that the fixed electrodes should be located away from anomalies if possible, to facilitate the spacing adjustment when they are re-positioned.

Electrode spacing It is evident from discussion above and, for instance, Fig. 18, that increasing electrode spacing will also increase depth penetration. However, this does not necessarily improve one's chances of detecting deeply buried features of modest dimensions, because, as Fig. 19 indicates, these will constitute an ever smaller part of the substantial volume encompassed by the current flow. The vulnerability even of shallow small features to increasing spacing is evident from Fig. 40.

Experiments were carried out in the electrolytic tank to ascertain the probe spacings giving optimum response with two of the more useful configurations – square and double dipole. In such idealized conditions, there are four variables: electrode spacing and the shape, size and depth of the feature. High- and low-resistivity square-section linear features were used. The results are shown in the approximate graphs (Fig. 44) in which the optimum electrode spacing is

Fig. 44. Optimization of electrode spacing for a square-section linear feature.

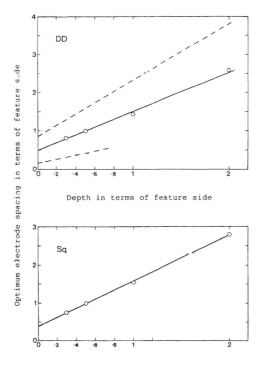

Depth in terms of feature side

plotted against the feature depth, both in terms of the length of the feature side (i.e. its width).

Let us use these graphs to derive a couple of practical examples. A typical Roman wall foundation might be 40 cm (16 in.) wide and 30 cm (12 in.) deep. The depth in terms of feature side is 0.67, which corresponds with electrode spacings of about 1.2 × feature side = 48 cm (i.e. about half a metre or 20 in.) for both configurations. For a ditch 2.5 m (8.2 ft) wide and again 30 cm deep, the depth in terms of feature side is 0.12, for which the optimum spacing is about 0.6 × feature side, which is 1.5 m (5 ft) for DD, and 0.5 × feature side = 1.25 m (4 ft) for Sq. These latter dimensions are approximately those for the Hog's Back barrow ditch. However, actual measurements for this, (Fig. 40) show better DD responses at 1 m (3.3 ft) and about 0.8 m (2.6 ft) than at 5 ft (about 1.5 m) spacing, indicating that the effective width of the ditch in resistivity terms is less than its actual width, as has been suggested on other evidence above. Clear-cut features such as walls are likely to behave more predictably.

The tank tests, supported by theory (Lynam, 1970), also showed that outside very approximate limits shown by the broken lines on the DD graph, a double-peak response is obtained. This would affect the Roman wall example if the electrode spacing were less than 20 cm (8 in.) or more than 1.9 m (6 ft) – rather extreme values. Extreme spacings can similarly affect the twin electrode configuration, but the problem is rarely significant in practice.

The ideal, in fact, tends to play a less than dominant part in the practical choice of probe spacing. One rarely has much prior information on the depth or scale of remains, so a good compromise between depth penetration and spatial resolution must be chosen. With the twin and square arrays, the spacing is arbitrarily defined by the fixed probe spacing of most survey frames. With the leapfrog method, the measurement spacing is tied to the reading interval and its choice may sometimes be affected by hardly scientific considerations of the area to be covered and the time available for the survey. The normal compromise is 1 m (3.3 ft). Before metrication, the most often used spacing in Britain was 1.22 m (4 ft) but this is highly incompatible with metric tapes! The slightly increased spatial resolution and a reduction of depth penetration which is not significant on the majority of sites, makes 1 m

(3.3 ft) a satisfactory new standard, at least for the initial evaluation of a site. With smaller spacings, not only is a survey slower, but probe positioning becomes more critical (see below). As we have seen, opening up probe spacing to increase penetration rapidly leads to loss of resolution and reduction of signal size, and the trade-off between these two factors must be carefully considered. In my experience, the widest spacing successfully used for a major survey was 2 m (6.6 ft), at Grime's Graves in Norfolk, an extensive Neolithic flint mine site (Sieveking *et al*, 1973). Initial tests showed that this was an appropriate spacing for the detection of the shafts, which are of the order of 12 m (40 ft) in diameter, and penetrating to the distinctively high resistivity rubble backfilling that they contain in their lower levels. It was decided that an increase in spacing might lose significant detail, even among these massive features.

Fig. 45. (A) Roman street in Winchester. Linear traverse in the Deanery garden revealed an impressive peak due to a Roman street, exactly 400 Roman feet from a known street. Alongside is a probable building. An uncomplicated plot for an urban site because it is on the fringe of the built-up area of *Venta Belgarum*, and of more recent development. (B) Traverses along two causeways crossing the ditch of Wansdyke (shown in schematic profile). (a) indicates that the ground has not been disturbed and the causeway is original; (b) shows that another causeway is built of loose rubble from the bank, and must be relatively modern. Soil is clay-with-flints on chalk, giving low background readings contrasting well with the chalk rubble.

Useful theoretical background for resistivity prospecting has been given by Aitken (1974) and his conclusions about the effects of length of probe insertion and errors of spacing, calculated for Wenner, are worth summarizing at this point: *Effect of probe depth*: for an increase from 5–15 cm (2–6 in.), less than 2 per cent for spacings in excess of 60 cm (2 ft). Probes with a depth stop will overcome even this small effect, which is practically insignificant anyway. *Probe positioning errors*: these are much more significant. With 60 cm (2 ft) spacing, errors of 2.5 cm (1 in.), or 4.2 cm (1.7 in.) for 1 m (3.3 ft) in placement along the line of the probes give reading errors of up to 4 per cent with the inner, potential, probes, and 3 per cent with the outer, current, probes. These are either additive or subtractive, depending on the direction of the error. The effect of a particular size of error rapidly diminishes with increasing spacing.

Errors due to lateral displacement of the electrodes from the straight line are relatively insignificant; therefore, if the insertion of a probe is impeded by a stone, it is always best to seek a new position at right angles to the line rather than along it.

Twin electrode and square array frames normally have a spacing of less than 1 m (3.3 ft) but, as this is fixed, positioning errors do not arise. As already mentioned, the standard 0.5 m (1.6 ft) for twin is closely equivalent in depth detection to 1 m Wenner. Hesse (1966) has made a study of the effect of temperature on ground resistivity, and

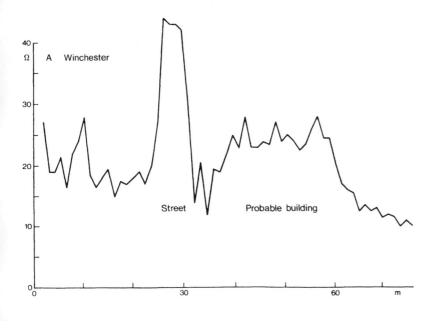

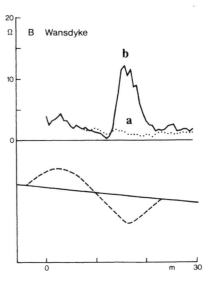

found it to be insignificant compared with water balance effects, except in such extreme conditions of freezing as have been discussed above.

Linear surveys versus area surveys Area surveys are ideal for the detailed planning of sites whose general location is known. Linear surveys, with separate traverses, can be effective as an economical way of searching a large area for buildings, roads or other substantial human artefacts, which can then be confirmed and planned with an area survey if desired.

A feature such as a Roman road or town wall, or a boundary ditch, can often be followed with well-spaced traverses 10–30 m (30–100 ft) apart so long as it is substantial enough to display a clear resistivity 'signature' and its likely course can be readily extrapolated (Fig. 45A). Single traverses can be very effective for discovering whether causeways across the ditches of ancient earthworks are original or not, and thus, for instance, for confirming the position of the entrance of an Iron Age hillfort (Fig. 45B).

In searching for an isolated feature such as a medium-sized building or a barrow ditch, the spacing will need to be kept down to 10–15 m (30–50 ft), and confirmatory traverses added if promising anomalies are located. Atkinson's approach to a ring ditch has been to run a traverse perpendicular to the line midway between any suspected ditch anomalies, to confirm the indications and establish the centre, and then to add one or two more radial traverses from the

Fig. 46. Ring ditch location. (1) Search traverse detects two likely anomalies. (2) A traverse at right angles halfway between these establishes the centre and diameter. (3 and 4) Two additional radial traverses confirm the ring ditch.

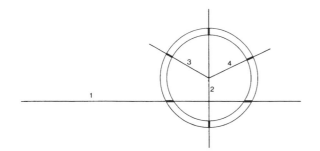

centre (Fig. 46). Another classic use of individual traverses was in locating the bastions of the Romano-British town of *Cunetio* (Fig. 5).

With more subtle features, credibility comes from repeatability. A major advantage of an area survey is that the large amount of information it contains reinforces the certainty of interpretation. Tell-tale patterns of plan emerge that cannot be shown by linear profiles, greatly increasing one's confidence in distinguishing man-made from natural features. Individual traverses may confuse geology with archaeology; they can also confuse one type of archaeology with another. At Winchester, during the excavation campaign of the 1960s, the direction of a major Iron Age defensive ditch suggested that it would pass beneath the open space called Oram's Arbour. A resistivity traverse revealed an appropriate anomaly exactly on the line predicted. Although other traverses were not as convincing, the evidence of the first was so impressive that they were dismissed as probably being affected by a different kind of filling – the wish was father to the ditch! Excavation showed that the convincing traverse had happened to be laid out over a medieval chalk pit, the outline of which would have been clear in an area survey. The Iron Age ditch had unexpectedly changed direction.

Resistivity in the future

Survey systems One of the dreams of practitioners of archaeological geophysics has been to develop a resistivity system that will give continuous traces in the manner of a fluxgate gradiometer. An apparatus designed by myself in 1972 was built by the School of Archaeological Sciences at Bradford University. An obvious principle to employ was the twin electrode, requiring only two moving contacts, combined with a constant current meter. Reminiscent of a lawn aerator, the device comprised two spiked wheels 0.5 m (1.6 ft) apart which acted as the electrodes, attached to a handle by which they were pushed. Spikes were chosen rather than discs because they were better able to make contact between stones, which tend to lift a continuous disc and break contact. This system was effective in not too adverse conditions but hard work to use, and motorization would have been a necessity (Fig. 47). The suitability of separate readings to the

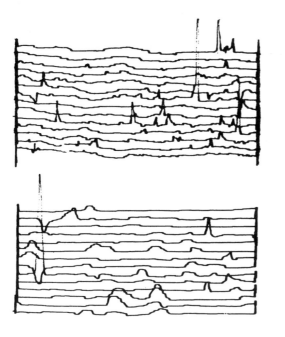

digitization of readings has deflected interest from such systems in recent times.

The French Centre de Recherches Géophysiques has developed a series of tractor-drawn assemblies using the square array configuration. One was a remarkable system using jets of ionized water as contacts, while another employed metal blades cutting into the ground. The latest such system uses a circuit with such high contact resistance tolerance that the electrodes need only to be wheels on the surface of the ground. These will operate even over light vegetation and contact problems occur only in the driest conditions.

In Britain, simpler systems are generally favoured on grounds of reliability and portability. As we have seen, great speed can be achieved by an experienced handler of the twin electrode frame with an automatic instrument and data logging. There is also potential for electrodes attached to the feet, with some sort of constraint to ensure that pace length is constant, although this has yet to be tried. These methods produce separate rather than continuous readings, but they are well suited to digitized logging and to the resolution normally obtainable with resistivity.

Another approach favoured in Britain is the use of multi-probe systems. In these, a complete row of probes is set into the ground, with individual leads back to a switching system linked to the instrument. Successive groups of probes are rapidly activated along the line. Apart from speed, the advantages of this system are that different probe configurations and spacings can be achieved simply by appropriate switching. Probably the first such system used was a manually operated one built by the British Ministry of Works Test Branch in the late 1950s. Since then, various versions using electronic switching have been developed.

Fig. 47. (Top) The continuously reading wheeled system. Two bladed wheels form a twin electrode array. (Middle) An assistant was needed to work the potentiometer distance transducer. (Bottom) A 30 × 14 m (100 × 46 ft) survey compared with a fluxgate gradiometer survey of the same area, below. Spikes on the magnetic survey are due to iron fragments; very similar spikes on the resistivity survey are caused by momentary loss of contact due to stones. Note that the magnetometer background level is flat because it is self-filtering, while the resistivity traces slope in response to a change across the area.

Normally, individual probes are used, and moved in succession from one line to the next in the wake of the measurements. This is more economical in movement than the leapfrog method, because each probe has only to be moved 1 m (3.3 ft) instead of 5 m (16 ft) for a 1 m (3.3 ft) spacing. Another approach is to use a bundle of cables with spikes at appropriate separations. This is simply laid on the ground and the spikes trodden in, and moved on to the next line after the measurements have been taken.

It is quite possible that all these ingenious ideas for speeding up resistivity may be overtaken by electromagnetic instruments, such as the EM38, coupled to an automatic recording system in the manner of the fluxgate gradiometer. But there remains something reassuring about the simplicity of setting up, lack of drift, and reasonable prices of conventional resistivity instruments which will ensure that they are never entirely superseded.

Vertical profiling Another much desired objective in all prospecting is to produce vertical profiles or cross-sectional pictures of subsurface features. To achieve this, the technique most readily appropriate on the archaeological scale is ground penetrating radar (Chapter 5), but encouraging developments are also occurring in resistivity.

A method developed by the OYO Corporation of Japan gives highly detailed vertical resistivity profiles, although it is limited to fairly stone-free situations. In alluvial deposits, it has proved effective for tracing the depth and extent of occupation-bearing layers first identified by trial trenching. Two ring electrodes 1 cm (0.4 in.) apart are set into a pointed probe of insulating polypropylene, 3.5 cm (1.4 in.) in diameter. These form one CP pair, while the other CP pair are normal probes set into the ground 10–15 m (30–50 ft) away on either side, forming a variant of the twin electrode configuration. The probe is driven into the ground by a hydraulic mechanism and its depth monitored by an opto-electronic transducer similar to that used in the Philpot magnetic recording system (Fig. 54). A combined resistivity meter and data logger (the Geologger) stores readings at 2 cm (0.8 in.) intervals on a computer disc, and these are subsequently displayed on a chart recorder. More will be said of this in Chapter 4.

Resistivity depth sounding can also be achieved by expanding the electrode spacing about a point, so that it is possible to obtain stratigraphic information in terms of the vertical distribution of resistivity. A reading obtained with, for instance, the Wenner array, is assumed to apply to the mid-point of the array at a depth of half the probe separation. Repeating such measurements along a line gives a crude cross-sectional picture or pseudo-section. This requires a large number of readings, and automated systems have been developed using multi-electrode arrays combined with microprocessors and electronic switching.

Griffiths and Barker at Birmingham University have developed such a system using 25 electrodes, with a control unit giving rapid readings and a completely flexible choice of configurations and spacings. A library of pseudo-sections representing different subsurface models has been built up, and computer matching programs are being developed. In a practical test, the rubble core of a hillfort rampart was convincingly outlined and tabular stone features have been distinguished from thicker structures.

Research is also under way to improve this type of representation by using the principles of tomography. Sheffield University has developed a medical imaging technique called Applied Potential Tomography (APT), analogous in function to the more familiar X-ray scanning tomography but using the principles of resistivity surveying to detect changes in tissue conductivity. The distribution of conductivity in an anatomical section is computed from voltages developed in an encircling array of electrodes, in response to currents passed between consecutive electrode pairs. The first practical scanner, developed at Sheffield, uses 16 skin electrodes to produce 104 independent voltage measurements from which the resistivity cross-section is computed in real time using a fast back-projection algorithm.

These principles can also be applied to a linear array of electrodes on a flat surface, such as the ground (Powell *et al*, 1987; Noël and Xu, 1991). With 16 electrodes, for example, there again exist 104 possible independent measurements of apparent resistivity, compared with only 35 measurements using the expanding Wenner configuration with the same electrode setting. All these measurements are gathered in order to obtain an

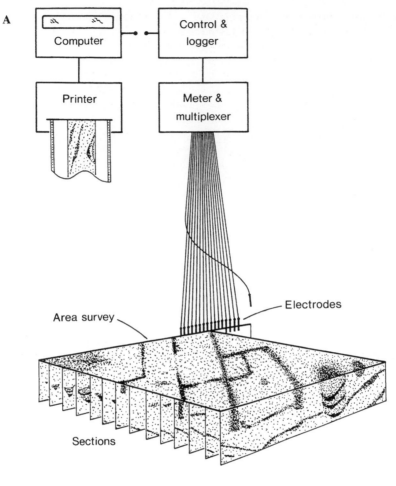

A

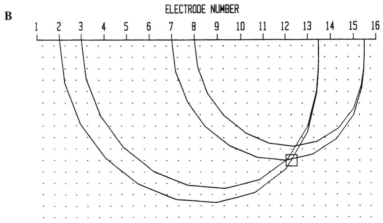

B

ELECTRODE NUMBER

1 2 3 4 5 6 7 8 9 10 11 12 13 14 15 16

Resistivity section

C

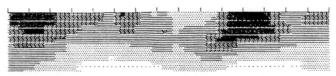

Fig. 48. (A) Resistivity tomography, schematic. A linear array of electrodes is used to make measurements in a variety of combinations determined by the control programming of the multiplexer switch. Measurements are stored in the data logger and computer processed on site to produce a series of resistivity sections and/or area maps at chosen depths. The archaeology can thus be visualized in three dimensions. (B) An idealized ground section. The potential gradient set up by current between electrodes 13–14 and 15–16 is sampled between 2–3 and 7–8, giving information about the resistivity of the pixels where the equipotential lines intersect. (C) Initial tests at *Verulamium* (St Albans), showing wall foundations. Electrode spacing is 0.75 m (2.5 ft).

optimum reconstruction of the resistivity section.

The potentials recorded at each step provide information on the resistivities of pixels which lie between equipotentials extending from the measurement electrodes (Fig. 48). The resistivity section is reconstructed, again using a back-projection method. The array is moved forward by leapfrogging the rear electrode so that by repeating the procedure a continuous image section is obtained. A practical result from initial tests at the Roman town of *Verulamium* is shown in Fig. 48C. The representation of wall foundations here seems quite convincing, and work continues to improve the speed and image-resolution of the technique.

In Japan, a different approach using DC resistivity has been successfully applied in archaeology by Imai, Sakayama and Kanemori (1987) of the OYO Corporation, using the method of alpha centres developed by Stefanescu and Stefanescu.

Although its effectiveness will be limited where the resistivity profile of a feature is not co-incident with its physical profile, as discussed earlier in this chapter, resistivity tomography offers the possibility of creating cross-sectional views of buried archaeology and also horizontal plans at chosen depths, both of which will assist in unravelling the stratigraphic sequencing which can appear as a baffling palimpsest in a simple survey plan. With the increasing power of portable computers, these advantages are becoming available immediately in the field and will be one further major step in the non-intrusive interpretation of archaeological sites.

Chapter 3
Magnetometry

Iron constitutes about 6 per cent of the Earth's crust, but little of it is readily apparent. Most of it is dispersed through soils, clays and rocks as chemical compounds which are very weakly magnetic. Man's activities in the past have redistributed some of these compounds and changed others into more magnetic forms, creating telltale patterns of anomalies in the Earth's magnetic field, invisible to a compass but detectable with sensitive magnetometers.

The geomagnetic field upon which these measurements depend seems to be caused by complex interactions between the Earth's hot, liquid, metal outer core as it rotates and convection within it, generating circular currents at the core-mantle boundary. These currents act as a solenoid to create the field, of which a distinctive characteristic is the varying angle of dip between the Poles and the Equator (Fig. 49).

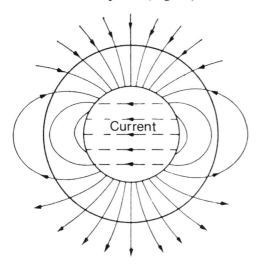

Fig. 49. The Earth's magnetic field, generated by an east-west flowing current regime at the core-mantle boundary. Adapted from Bott (1971).

Principles

All magnetism is ultimately due to the movement of electrical charge. Materials become magnetized by the rotation and spin of the negatively charged electrons which orbit around atomic nuclei. The magnetic behaviour of matter takes a number of different forms, depending on whether the crystal lattice permits the magnetic fields of electrons to reinforce or oppose each other, and in what degree. Ferromagnetic substances such as iron are in the first category, but the compounds most significant in archaeological prospecting are the oxides magnetite, haematite and maghaemite. In these, the electron fields are partly balanced out, so that they are relatively weakly magnetic. They can retain a permanent, or remanent, magnetization when placed in a magnetic field and, because of their magnetic susceptibility, can also acquire a magnetization in the presence of a field; this is lost when the field is removed, and is called temporary magnetization. The various magnetic states will be discussed more fully in Chapter 4.

Units of measurement The unit of magnetic flux density used here is the SI sub-unit nanotesla $(nT) = 10^{-9}$ tesla (T), which is numerically identical with the formerly used gamma sub-unit of the cgs system. Gamma is still often used informally because it is easier to say. Magnetic susceptibility units will be discussed in Chapter 4.

Thermoremanence As we have already seen, the magnetism of baked clay is relatively strong and has long been appreciated. The compounds initially present in the raw clay include those mentioned above and others that are very weakly magnetic, and all have randomly orientated fields producing little net effect. The firing process is an efficient way of achieving magnetic enhancement because the weakly magnetic compounds are

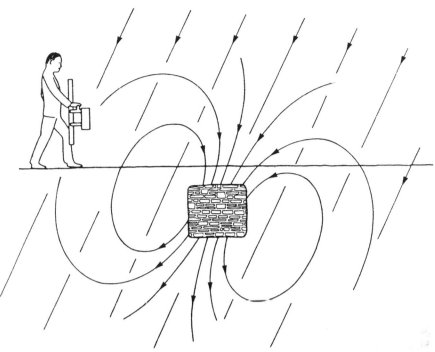

Fig. 50. The field of a thermoremanently magnetized kiln in mid-northern latitudes. Note the angle of dip, which causes the maximum of the anomaly to be displaced to the south. The fluxgate gradiometer outlined is the early Plessey 1 m (3.3 ft) type.

converted to oxides including magnetite, one of the more strongly magnetic iron compounds. Then all the oxides are completely demagnetized at the Curie point (675°C for haematite, 565°C for magnetite) and, on cooling, are re-magnetized *en masse* by the Earth's field so that a feature such as a pottery kiln or a clay hearth ends up with a fixed and fairly coherent permanent magnetization aligned with the geomagnetic field at the time of firing (Fig. 50).

The concept of 'relatively strong' needs to be kept in perspective. A pottery kiln can sometimes give a signal as great as 500 nT near ground level whereas some archaeological anomalies are less than 1 nT; but even 500 nT is only about one hundredth of the present geomagnetic field intensity in Britain.

Stones and soils can also acquire thermoremanence, both sometimes as strongly as clay if they have a reasonably high magnetic mineral content and the net effect has not been randomized by subsequent disturbance. Basalt and other igneous rocks are especially magnetic, having been thermoremanently magnetized when they initially cooled.

Magnetic susceptibility This phenomenon has already been alluded to above. Chapter 4 is devoted to it, but it is so important in magnetometer surveying that it must be introduced here. Thermoremanence is effectively permanent, but a material with magnetic susceptibility is only magnetic in the presence of a magnetizing field – although in our case the Earth's field is always present, so there is no danger of the induced magnetization disappearing! Both types of magnetization look the same to a magnetometer and there is rarely any observable practical difference.

Variations in magnetic susceptibility between topsoils, subsoils and rocks also affect the Earth's field locally, making it possible to detect ditches, pits and other silted-up excavated features, and even buried walls; it has produced some of the most subtle effects in magnetic prospecting. Its great value in archaeology derives from two factors. First, topsoil is normally more magnetic than the underlying subsoil or bedrock, so that excavated features silted or backfilled with topsoil will produce a positive magnetic signal; conversely, less magnetic material intruding into the topsoil, including many kinds of masonry, can be detectable by a subtractive effect which gives a negative signal. Second, magnetic susceptibility is particularly subject to enhancement by human activities, so that magnetic prospecting in its different forms has the great power of being selectively sensitive to the former presence of human occupation.

When considering a survey, other than of normally reliable thermoremanent features such as kilns, it is essential to ascertain whether there is sufficient contrast in magnetic susceptibility between topsoil and natural for features such as ditches to be detected – even the clay of kilns can be weak in some iron-deficient areas. High susceptibility in topsoil alone is not enough: if the natural is just as magnetic, there will be no detectable contrast. Preliminary susceptibility measurements are highly desirable, but if the equipment is not available a scan can suffice to ascertain whether anomalies are detectable. Tite has already published data for Britain and abroad, and a large number of soil susceptibility measurements has been collected by the Ancient Monuments Laboratory for the construction of a 'surveyability' map of Britain. Until this appears, a preliminary guide is offered below. This can only provide a broad indication of the likely effectiveness of surveys. There are many local variations and the best general rule is to 'have a go', however unpromising the geology appears to be, especially as the sensitivity of instruments and plotting methods continues to push back the limits of archaeological detection and 'sub-nanotesla' surveys can be routinely accomplished.

Instruments

Proton instruments This type of instrument was the first to make magnetic prospecting practicable in archaeology, and its elegant principle has several inherent advantages: readings are absolute, requiring no calibration; it measures total field without any direction sensitivity; and it requires no setting up procedure, high-precision construction or rigid support (Packard and Varian, 1954; Waters and Francis, 1958). But, in spite of this catalogue of virtues, proton instruments have been largely displaced because their two-stage polarization-measurement mode of operation makes them slower than new instruments now available, and less well adapted to the rapid gathering of high-density information now demanded of archaeological surveys.

Protons are the nuclei of hydrogen atoms, of which there are two in every molecule of water. Because they are spinning, the protons are magnetized, and behave as very small bar magnets, or magnetic dipoles.

The principle of the proton free precession detector is illustrated by the example shown in Fig. 51 (Hall, 1962). It consists of a 250 ml polythene bottle containing a proton-rich liquid. Water is highly effective, but if there is any danger of freezing alcohol is preferable. (Methanol – methyl alcohol – is shown.) The bottle is surrounded by a coil of copper wire and the whole is encapsulated in epoxy resin to protect it. Its operation is a two-stage process of polarization followed by measurement. The diagram shows the detector at the polarization stage, in which a DC current of about 1 amp is passed through the coil, so that it acts as a solenoid or electromagnet which tends to align the protons parallel with its axis. The current is then switched off and the coil becomes a detector connected to a sensitive amplifier. The protons turn to align with the ambient magnetic field, which is that of the Earth plus any magnetic anomalies. Because they are spinning, the protons precess like falling tops as they realign, generating a small alternating (AC) voltage in the coil. The frequency of this, which is measured, is exactly proportional to the strength of the field. The bottle is aligned roughly east-west to maximize the angle through which the protons re-orientate, and therefore the amplitude of the signal.

Toroidal detectors have also been constructed in which the coil is wound around a doughnut-shaped liquid container, and these are claimed to reduce interference. A very steep magnetic gradient can spoil the operation of detectors because it will cause the protons to precess at different rates, so that a coherent frequency cannot be measured. When this happens, the signal is said to be 'killed'.

Single detector measurements are unsatisfactory for sensitive work because they are affected by background variations and interference of different kinds. The most significant natural effect is the diurnal variation due to the interaction of the

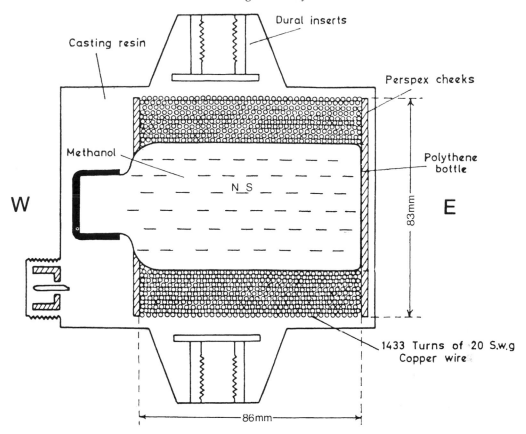

Fig. 51. Casting resin, Dural inserts, Perspex cheeks, Methanol, N S, W, E, Polythene bottle, 83mm, 1433 Turns of 20 S.w.g Copper wire, 86mm

Fig. 51. Design of a proton magnetometer detector. The protons are shown symbolically as small bar magnets. Adapted from Hall (1962).

Earth's field with the solar wind, the flux of charged particles from the sun. As the Earth rotates, this causes a rapid fall in the geomagnetic field intensity during the early morning and a corresponding increase toward the end of the day. The overall variation in Britain can be as little as 5 nT in winter, but 50 nT in summer. Larger variations can be caused by magnetic storms, especially during periods of high sunspot activity. Earth currents from DC electric trains cause severe interference for up to 16 km (10 miles), which would seem, for instance, to rule out the whole county of Surrey! A single detector is also as sensitive to interference at and above ground level as it is to magnetic features below the ground, so that wire fences, structural iron in buildings and passing traffic can all be troublesome.

These problems can be dramatically diminished by using differential measurement. In this,

the difference in signal between two identical detectors is measured, on the principle that interference will affect both detectors equally, producing no net effect on the reading. The second, compensatory detector is known as the reference detector. The preferred modern approach is to keep the reference detector in a fixed position outside the survey area; this gives the full sensitivity of a single detector and provides excellent compensation for diurnal variations, but is less effective for local interference near one detector or the other. Modern differential instruments measure the frequency of both detectors, sometimes using two separate instruments, and simply display the difference. A resolution of 0.1 nT is achievable. There are a number of suitable instruments on the market, mostly from the USA. For the best results, it is important to choose one that makes use of simultaneous, rather than sequential, measurements from the two sensors as the latter can be affected by micropulsations – very rapid small changes in the geomagnetic field.

The alternative is to mount the detectors on a single support staff, as a gradiometer (Fig. 10), with the reference detector above the measurement detector (Aitken and Tite, 1962). There is some loss of sensitivity with this arrangement, because the reference detector is near enough to be somewhat affected by buried features, and the difference signal is thus reduced. It does, however, give a degree of 'self-filtering', in favour of features beneath the lower detector, while local interference above ground level is minimized because it tends to affect both detectors equally.

A gradiometer with simple circuitry can be made by sending the polarizing current through both detectors in series, and then, by means of a switch, sending their combined signals to an amplifier whose output is passed to a loudspeaker or meter. The fundamental frequency is about 2,000 Hz in medium latitudes, nicely within the audible region. A difference in precession frequency between the detectors will appear as beats as the two waves alternately reinforce and subtract from each other. The beat frequency is calibrated in terms of the signal difference between the sensors – the faster the beats, the bigger the difference. A convenient rough-and-ready timer is the duration of the precession signal, which disappears in two to three seconds, so that one can simply count the number of beats before the signal dies away. This type of instrument is within the capabilities of an amateur to construct, but its resolution is not high and it cannot discriminate between positive and negative signals without some additional circuitry, for instance a small solenoid between the sensors to create a field of known sense which will show the sense of the measured signal by either reducing or reinforcing it. For modern large surveys, it is too demanding and slow to use.

Alkali vapour magnetometers This is a highly sensitive type of instrument, pioneered in archaeology in the 1960s by the University Museum of Pennsylvania University and Varian Associates (Ralph, 1964). They are also known as optically pumped, optical absorption or caesium or rubidium magnetometers, depending upon the element used. They have the advantage of producing an effectively continuous signal and can be used in differential form, but they are expensive and rather prone to breakdown because

of their complexity. For everyday work, their high resolution of about 0.01 nT rarely has a practical advantage over fluxgate and proton instruments of 0.1 nT resolution because of limitations set by 'soil noise' – background signal irregularities due to variations in the soil. However, as we have seen, a caesium magnetometer was of critical value in the search for the Greek town of Sybaris in Southern Italy, where non-magnetic stone remains were deeply buried in a weakly and quite uniformly magnetic alluvium.

The principle is sophisticated, analogous to the proton magnetometer but more complex. It operates at the atomic rather than the nuclear level and a lamp is used for polarization, by the process known as optical pumping, and also for monitoring the signal. Let us consider the caesium version. The sensor is a glass cell containing caesium vapour at low pressure. Light from a caesium vapour lamp, after filtering and polarization, is passed through a glass bulb also containing caesium vapour, at low pressure, and then to a photocell. The light excites electrons in the caesium atoms to a more energetic state, losing energy and becoming dimmer in the process, and the photocell detects this. The electrons quickly fall back to their original energy level, but are then re-excited. As the process continues, the magnetic vectors of the atoms tend to align with the light beam, with a reduction in the light absorbed, but they also tend to realign with the ambient magnetic field, precessing as they do so. An alternating magnetic field is applied to the glass cell by means of a coil around it, and the frequency adjusted until the light passing through goes to a minimum. The frequency of the field is then equal to the rate of precession and reinforcing it, and more light energy is required to re-polarize the atoms.

For practical application, another effect is normally used. The precession also has a small modulating effect on the intensity of the light beam, and this can be monitored to activate a self-oscillator which continuously follows the precession frequency.

Because of various complex interactions between the ambient magnetic field and the precession process, the instrument is direction-sensitive, and needs to be maintained within 5 degrees in direction (heading) to keep the orientation error within 0.1 nT, although this can be improved by using a gradiometer configuration.

Also, although the sensitivity is very high, the absolute accuracy is not. The sensitivity of the instrument derives from its high precession frequencies, which make small signals much easier to measure than with the proton principle.

Fluxgate gradiometers At one time it was assumed that for all normal archaeological prospecting a proton magnetometer should be used, and for the highest sensitivity an alkali vapour instrument. The fluxgate gradiometer, with its closely-spaced direction-responsive detectors, was regarded as a conveniently fast but relatively insensitive device, prone to heading problems and drift. Continued development and appropriate methods of use have solved these problems and the fluxgate gradiometer has become the workhorse – and the racehorse – of British archaeological prospecting.

The action of the fluxgate sensor is shown in Fig. 52. The high magnetic permeability cores are normally strips of an alloy called Mumetal. They are driven in and out of magnetic saturation by the solenoid effect of an alternating 'drive current' in the coils wound round them, normally with a frequency of about 1,000 Hz. Every time they come out of saturation, external fields can enter them, causing an electrical pulse in the detector coil proportional to the field strength. Two Mumetal strips are used, with the drive coils in opposite directions – series-opposition – so that the drive current has no net magnetic effect on the sensor coil. A typical fluxgate element is about 50 mm (2 in.) long and 5 mm (0.2 in.) in diameter.

Although the function of the fluxgate involves a cycle of operations, as does the proton sensor, these take place in one-thousandth of a second rather than about five seconds, so that the output is effectively continuous. Another difference is that the proton sensor is omni-directional, measuring total field, whereas the fluxgate is highly directional, measuring only the component of the field parallel to its axis, i.e. along its length. This is an advantage in some applications, but a definite disadvantage in archaeological survey: with a

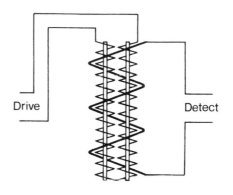

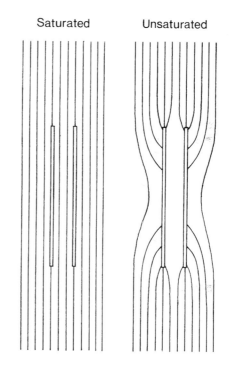

Saturated   Unsaturated

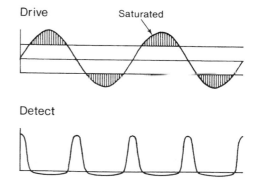

Drive   Saturated

Detect

Fig. 52. (Top) Schematic diagram of fluxgate detector. (Centre) Cores in magnetic field. When they are unsaturated, the field enters them, cutting the detector coil and generating a voltage in it. (Bottom) Sine wave drive input current and detector coil output voltage.

single sensor, the slightest tilt causes the amount of the ambient field flux along the axis to change, with a consequent change in reading. The problem is solved by using two sensors arranged as a gradiometer, with the output of one subtracted from the other. If they are accurately aligned and matched in sensitivity, no signal is detected in uniform field, whatever the angle of the pair in the field. Thus the gradiometer configuration, which we have seen to be desirable with the proton instrument, is essential with the fluxgate type if its capability for rapid measurement is to be utilized. Instruments with a vertically suspended single detector do exist for geological applications but they are slow to use; three-element detectors are constructed for total field measurement, at considerable increase in cost and complexity.

The practical effect of any misalignment of the detectors is to make the instrument direction-sensitive, so that the zero changes if it is rotated. To achieve the necessary precision and stability of detector alignment, fluxgate gradiometers have to be constructed with great mechanical care and rigidity. In earlier instruments with 1 m (3.3 ft) detector separation this was quite difficult, but the necessary stiffness is much easier to achieve with the 0.5 m (1.6 ft) separation now general. The stability of the shorter instrument outweighs the greater sensitivity of the longer one. The difficulty of maintaining the stiffness of the structure, and thus achieving a low 'noise' level, increases rapidly with length and the instrument also becomes more cumbersome. It thus became clear that the way forward was to use the shorter instrument, and to concentrate on improvements in electronic noise level.

The approach of Philpot has been to mount the detectors in a stiff quartz tube protected by an outer plastic tube, the mountings of the inner tube being positioned for minimal bending of the tube, whatever its attitude. The Geoscan FM series relies on the stiffness of a wider square-sectioned aluminium tube. Even so, controls are needed to trim the detector alignment finely, and thus to reduce direction sensitivity to a minimum. This is done by first pointing the instrument north-south, then east-west, and adjusting the control on the north-south line through the instrument in each case. The process is repeated until direction sensitivity is minimized.

Both instruments have controls and electronics in a compact housing conveniently joined to the tube by a handle near its top. The control panels of two versions of the FM instruments are shown in Fig. 53.

Fig. 53. Control panels of two Geoscan fluxgate gradiometers. The analogue display (top) provides the most suitable visual output for scanning. The FM36 (bottom) has comprehensive controls for the built-in data logger, and the digital display includes the grid, line and measuring point numbers.

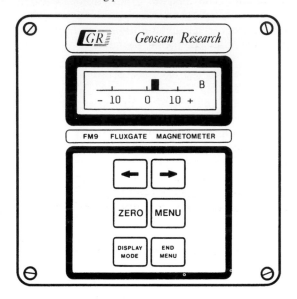

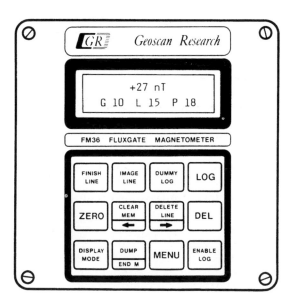

The Geoscan instruments have a noise level of about 0.1 nanotesla, so that 'sub-nanotesla' surveys in areas of weak magnetic contrasts are now readily achievable. Thus the practical sensitivity of fluxgate instruments is close to that of the best proton instruments and to the generally realizable limits of optically pumped instruments, with the additional advantages of compactness, relative simplicity and cheapness combined with suitability for high resolution automatic recording.

Survey methods and recording systems

Grid marking Naturally enough, the first method of recording was to write down the instrument readings. Aitken used an 'instant grid', made from a net of plastic coated clothes line (Fig. 11), about 50 ft (15 m) square, and with a mesh of 3 m (10 ft). This was hooked on to successive squares of a previously laid out grid of pegs, and moved on by two people at the corners, dragging one edge across the rest of the net. Instrument readings were made at 5 ft (1.5 m) intervals by interpolating within the net. It was simple and quick to use in open areas, but impossible among bushes and trees. It was also sometimes extremely difficult to untangle and had to be constantly guarded in the presence of sheep, by whom it was regarded as a rare delicacy.

The 5 ft (1.5 m) spacing between readings enabled surveys to be manually recorded and plotted with reasonable speed, and was suitable for defining major features such as ditches and kilns. But spatial resolution was poor. A strong localized anomaly had to be checked out by taking additional readings around it to ascertain whether it was likely to originate from archaeology or superficial iron, a process which slowed the survey down. This type of net was used with an Elsec proton magnetometer by the Ministry of Public Building and Works Test Branch, but they supplemented it with a 25 ft (7.6 m) net with a 5 ft (1.5 m) mesh, enabling readings to be taken at 2.5 ft (0.76 m) spacing for detailed work (Fig. 12). More recently, a reading interval of 1 m (3.3 ft) has become standard, and is a reasonable compromise that will give an adequate general impression of the density of occupation of a site without the necessity for intermediate readings.

More convenient successors to nets are separate plastic coated lines, marked at suitable intervals with adhesive tape, which are really nets broken down into more manageable elements. The use of these will be described in Chapter 8.

Fluxgate gradiometers and continuous recording The first steps in this important development have been described in Chapter 1.

The system as finally evolved by the Ancient Monuments Laboratory made use of a 0.5 m (1.5 ft) Philpot instrument combined with a Servoscribe M battery-powered XY plotter (Fig. 54). The instrument signal is shown on the vertical Y-axis, successive traverses being separated in proportion to their separation in the field by means of a stepping potentiometer signal added to that of the instrument. The distance signal, moving the pen proportionally along the X-axis, comes from a potentiometer driven by a string and pulley arrangement supported by two non-magnetic tripods and moved by the instrument carrier. It is easy to grip the string in one hand and carry the instrument in the other so that it is 0.5 m (1.5 ft) from the string; thus, by walking up one side and down the other (and incidentally using the string as a guide), two traverses 1 m (3.3 ft) apart are recorded. A marker on the string is stopped over the two tapes which define the two sides of the survey square between which the string is stretched. The tripods are moved on in 2 m (6 ft) steps after recording each pair of traverses, the instrument carrier moving one and an assistant (who also looks after the recorder and trace stepping) the other. The traces were originally stepped automatically by switches on the tripods actuated by small buttons on the string, but manual stepping at the recorder end proved to be more reliable. To avoid the effect of any residual direction sensitivity in the fluxgate gradiometer, its heading is kept constant for both traverse directions. The basic logistics of this method of surveying are still widely used, as described in detail in Chapter 8.

Some technical details and solutions learned from experience and experimentation are worth adding, especially as some modern computer-recording systems are based on the same principle. A perfectly adequate potentiometer is an ordinary small ten-turn type (for example, from RS Components), suitably geared to a flat-rimmed pulley wheel, which can simply be a plastic trolley wheel obtainable from a DIY shop. A

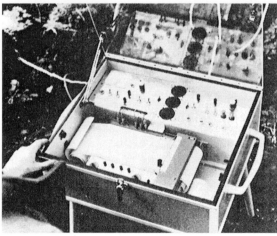

Fig. 54. Evolution of automatic fluxgate recording systems. (Top) The first Ancient Monuments Laboratory version, with a Plessey gradiometer, at Briar Hill, Northampton. Note the pulley for the distance transducer string on the tripod. (Bottom left) The recorder and controls. (Bottom right) The later lightweight version at Avebury. (Top opposite) The Philpot system, the first with computer recording, in Oman. A single string is wrapped round a hand-held pulley driving an opto-electronic distance transducer. (Bottom opposite) Modern computer display of data from instrument logger (resistivity survey shown).

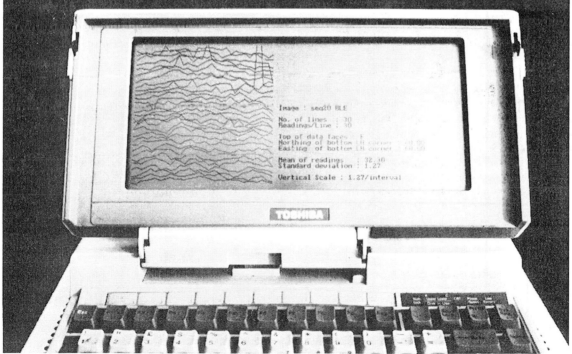

convenient diameter is 5 in (12.7 cm. The tyre is removed and replaced by $^{1}/_{4}$ in (6.4 mm) rubber bands placed side-by-side to give a non-slip surface for the string. It needs a couple of guides, which can be loops of thick stainless steel wire (hard and preferably non-magnetic) to keep it on the pulley. The string is medium thickness (about 1 mm) nylon monofilament, as used for fishing lines. This has the virtues, compared with woven nylon string, for instance, of having a small cross-section so that it is not too much deflected by wind, and of not being too prone to stretching: after some initial stretching when first used, it seems to work-harden rather like a metal and becomes quite stable. The pulley on the second tripod acts simply as an 'idler' to support the string; almost any small dished-rim plastic wheel, again with its tyre removed, will suffice for this.

Needless to say, the metals used for this system should be non-magnetic, but if any magnetic components are unavoidable their influence on the instrument can be minimized by keeping the tripods well back from the survey area. A simple gear train is needed for linking the pulley to the potentiometer. Tripods must be foldable, telescopic or demountable; they can be specially made for instance from aluminium angle, but robust camera tripods with added horizontal struts are very convenient and transportable. The tripods need to be weighted to prevent them being pulled over and, to save weight in transportation, this is conveniently done by attaching a stone or brick from the site to a horizontal cross-member between the rear legs, made of non-magnetic angle suitable for holding the weight.

A great advantage of this method is that very high spatial resolution data (Fig. 55) are recorded automatically while the instrument is carried at a normal walking pace. This resolution is limited only by the response time of the system and any mechanical backlash in it. The detailed traces

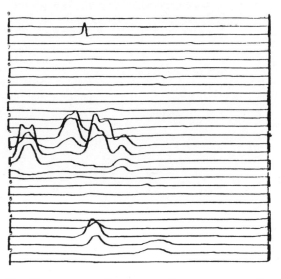

Fig. 56. Automatic fluxgate gradiometer survey plot of a standard 30 m (100 ft) square, showing three Roman pottery kilns, left of centre, with characteristic double peaks caused by their walls. Sensitivity: 50 nT/line interval.

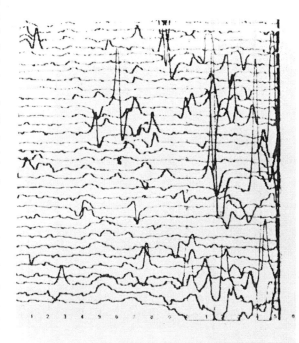

Fig. 55. Automatically plotted 30 m (100 ft) square, part of the survey of the Roman settlement at Neatham, Hants. The line of small peaks from top to bottom, left of centre, clearly reveals a very small ditch in the presence of spectacular interference from modern iron rubbish. It would have been difficult, or impossible, to understand what was going on in this square without high resolution, closely spaced traces running in this direction. Sensitivity: 7.5 nT/line interval.

Fig. 57. (Top) Roman pottery kiln in good condition, with oven floor preserved. This would give a single peak. (Bottom) A kiln in more usual condition, with oven floor missing, gives double peaks as in Fig. 56.

reveal the characteristic signatures of feature types, facilitating their recognition and greatly improving the overall standard and confidence of interpretation – in contrast with separate readings which can be likened to glimpsing the anomalies through small holes in an opaque screen. Not the least significant advantage of this resolution is that spikes in the traces due to relatively superficial modern iron rubbish are immediately distinguishable from archaeological features, which are therefore traceable in conditions of interference that would have made interpretation impossible with normally spaced separate readings (Fig. 55).

However, high resolution is achieved only in one direction. At right angles to this, the traverses are separated by 1 m (3.3 ft), and the resolution is no better than with separate readings recorded at this interval. The poor resolution between traverses can make it difficult even to recognize linear features, such as ditches, parallel to the traces, let alone to analyse them in detail (Fig. 58). If time is available, the problem can be remedied by surveying the area a second time at right angles, or by halving the traverse spacing – although the latter approach can sometimes produce confusing results. Modern digitized measurements are less affected by this problem, as will be seen below. Another important advantage lay in the immediacy of the record obtained. This made it possible to monitor progress, and to know when detectable features had ceased and the survey could be stopped – and to give archaeologists plots on site in urgent situations.

The survey in Fig. 56 was one of the earliest made with the system, and impressively confirmed its ability to reveal the nature of buried features by their characteristic signatures. One of the central group of peaks has two clear minor peaks superimposed on either side of it, while adjacent peaks above and below are slighter and less complicated. Other peaks also show this effect, though less clearly. This response pattern was interpreted as being due to the circular fired clay walls of kilns that had lost their superstructures (Fig. 57), and they did indeed prove to be Romano-British pottery kilns in this condition. Resolution as good as this is only achievable with fairly shallow features, from just below the ploughsoil to about 0.5 m (1.6 ft) deep. The effect of increasing depth is to smooth out the detail in the anomalies.

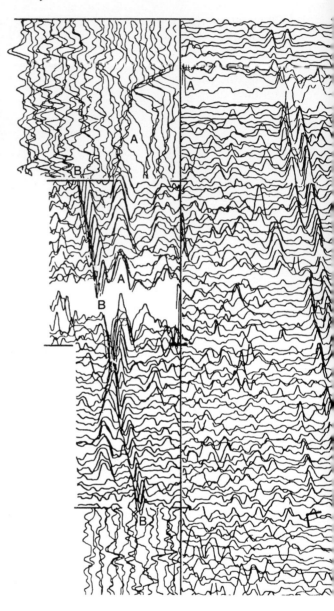

Fig. 58. Llawhaden, Dyfed. Limitations of trace plotting. Part of a strongly magnetic site surveyed in different directions. The large angled ditch, A, remains recognizable in spite of great changes in aspect, but the narrow linear feature, B, disappears when nearly parallel with the traces. Sensitivity: 7.5 nT/line interval.

Other anomalies in the square are an 'iron spike' and weaker, uncomplicated peaks indicative of a pit. These do not show more strongly because the sensitivity of the system was reduced to show the strongly magnetic kilns clearly. Herein lies one of the limitations of the system: the inflexibility of its sensitivity. The recording sensitivity must be set at the beginning of a survey and is established on the basis of available evidence: a preliminary scan with the magnetometer; knowledge of the sensitivity to be expected on the site lithology; and the nature of the expected remains, usually ranging from slight ditches to heavily fired clay structures. To produce a presentable picture, the sensitivity is then normally kept constant throughout the survey, but problems are encountered with sites such as this where a low sensitivity is needed for the kilns and a higher sensitivity to show associated features. The nearby pit is strong enough to show up but other, even weaker, pits and ditches might have existed – and indeed were found in a re-run of the survey at high sensitivity. Thus, in such a case as this the survey ideally needs to be made twice, at two sensitivities, to extract as much information as possible.

Another major problem deriving from the inflexibility of the direct recording system is its inability to translate the data into any other form of presentation, such as dot density. This was overcome with some early surveys by using a manual trace follower to digitize the trace data for computer processing. This was prohibitively slow and labour-intensive, especially with old types of digitizer, and complex surveys with overlapping traces could be almost impossible to sort out. Normally, all that can be done routinely is to produce a reduced scale montage of the traces as a final picture. Although full of detail for the skilled interpreter, such plots, unless very clear, tend not to be readily acceptable to the archaeologist or the general reader. One problem in visualization is caused by the superimposition, in the vertical direction, of response signals and positional signals so that anomalies, especially strong ones, seem to be displaced from their true positions. Data-logging in parallel with the trace-plotting was tried, but little was achieved in this direction until the appearance of miniature portable computers in the early 1980s – with the notable exception of the pioneering work of Sowerbutts mentioned in Chapter 1.

Computerized fluxgate systems These have evolved by progressive modification and adaptation of the direct recording principle.

The Philpot system includes a distance transducer consisting of a single string stretched between tripods. The string is wrapped once round a pulley which is hand-held by the person carrying the instrument and drives an optical encoder (Fig. 54). The distance and instrument signals are sent in digital form by a single cable to an Epson HX20, or other portable computer, which stores the data on tape. The HX20 will produce a miniature trace plot built up from dots, which is adequate for monitoring survey progress. The system is programmed to collect data at suitable intervals, typically 0.3 m (1 ft), and to recognize automatically when a full traverse or line of readings has been recorded.

The Geoscan system is available in various alternative forms. The basic instrument, the FM9, is for use either without a data logging system or with an external one. The FM18 and FM36 have built-in loggers, with capacities of 4,000 and 16,000 readings respectively. Data collection can be initiated by a manual switch carried by the operator as he passes points on a tape or marked string, or automatically with a string-and-pulley switching system. Survey progress is monitored on the HX20 by means of dot-density plots.

A remarkable new Geoscan approach, based on an idea of Lewis Somers, is the ST1 sample trigger used with FM18 or FM36 gradiometers. The function of the ST1 is to sound a 'bleep' every metre (3.3 ft), while triggering the gradiometer to record readings in its internal data logger at a selectable rate of 1, 2, 4 or 8 per metre. Using a string or strings marked in metres laid across the grid square, the operator has to make his arrival at each metre point coincide with a bleep, a skill that is readily acquired, the more easily if he can adjust his pace to 1 m (3.3 ft). Distance errors appear to be within 0.25 m (10 in.), which is generally acceptable, and should not be cumulative, providing the operator is keeping up, because the data logging process is reset for each traverse. Line lengths are selectable at 10, 20, 30 and 50 m (33, 66, 100 and 160 ft). As with conventional surveys, it is convenient to walk along either side of a string at a distance of 0.5 m (1.5 ft), although experienced operators can dispense with the string entirely (see Chapter 8). The method seems best suited to relatively smooth ground, over which it produces surveys of high quality; there can be difficulties in keeping up with the bleeps in ploughed fields on heavy soils (see Addendum, p.171).

The ST1 and the Somers and Bartlett variation, mentioned below, have made possible the fastest surveying speeds yet achieved in archaeological prospecting. In 1989 the records stand at a 20 m (66 ft) square in 5 ¼ minutes, and a 30 m (100 ft) square in 12 ½ minutes, compared with about 20 minutes for a 30 m (100 ft) square with the XY plotter system. So far the impressiveness of this speed has been diminished by the slowness of data transfer from the data logger to a computer, which normally has to be done after each square. But this problem is being overcome by improvements in both portable computers and programming. It is circumvented by a system devised by Somers and Bartlett which retains the cable between instrument and computer so that reading and final data storage are simultaneous.

The quality of the surveys achievable by this method is in no sense diminished by the speed; rather the close detail revealed over wide areas, combined with modern computer graphics, produces images of breadth and subtlety. The laying-out of the grid is becoming the slowest part of a magnetic survey. The Epson HX20 was the first truly portable computer and initiated the recording revolution, but its screen and printer are small, and its speed of data processing slow. The Ancient Monuments Laboratory therefore now uses the Geoscan magnetometers, as well as resistivity equipment, with a Toshiba T1200 portable computer with hard disc storage. This has a large screen which displays a square of data at a scale similar to that of the old direct plotter paper, with the elegant and much appreciated advantage that it does not blow away in the wind or become pulp in the rain. Even with this, data transfer still takes significant time, but improvements are under way.

Sensitivity and resolution

In all archaeological prospecting work, as we have seen already in Chapter 2, there are two factors that one strives to maximize – depth sensitivity and spatial, or horizontal, resolution – yet these are difficult to reconcile. In both resistivity and magnetometer surveys, signals tend to broaden as well as to weaken with depth, and clarity of

definition falls off rapidly below a depth of about 1 m (3.3 ft).

Depth sensitivity When considering the strength of signal from a buried feature, a useful approximation can be made by assuming that all its magnetism is concentrated in a dipole aligned with the geomagnetic field at its centre of mass, or centroid. In a discrete feature such as a cylindrical pit, this point will be at its geometrical centre, while in an ideal triangular-section ditch it will be at the intersection of the medians – lines from the vertices of the triangle to the mid-points of the sides (Fig. 61). For the discrete feature, it can be assumed with fair accuracy that the strength of its magnetic anomaly falls off inversely as the cube of its depth below the detector, and for the linear feature as the square of the depth (Linington, 1964).

Fig. 59 compares the fall-off with depth in British latitudes for a total field instrument such as a single sensor proton magnetometer (or a differential one with a remote reference detector), and a 0.5 m (1.6 ft) fluxgate gradiometer. The bands are delimited by the discrete feature and linear feature responses; archaeological features with shapes somewhere in between should lie within the bands. The assumption is made that the absolute instrument gives a response of 1 at a depth of 1 m (3.3 ft), a reasonable mean depth for a large proportion of the archaeology of Britain and elsewhere. At this depth, the fluxgate instrument is only 0.52 to 0.65 times as sensitive, while at greater depths its relative sensitivity drops lower but not dramatically so; overall, the loss in sensitivity for practical purposes is about a half. The sensitivity of a 1 m (3.3 ft) fluxgate gradiometer lies about halfway between, at about three-quarters, but such longer instruments present problems already alluded to.

Some further points emerge from Fig. 59. There is a rapid fall-off in sensitivity of both instrument types between 1 and 2 m (3.3 and 6.6 ft), and by 3 m (10 ft) the limit of detection is effectively reached for most features. At 3 m (10 ft), a kiln giving a good average anomaly of 200 nT at 1 m (3.3 ft) with the proton instrument (130 nT with the gradiometer), would give only 7.4 nT (2.6 nT). A pit giving 8 nT at 1 m (3.3 ft) would be reduced to 0.3 nT (0.1 nT) and would be undetectable above the soil and instrument 'noise'. The only significantly better response at depth is that of the proton instrument to a linear feature, so

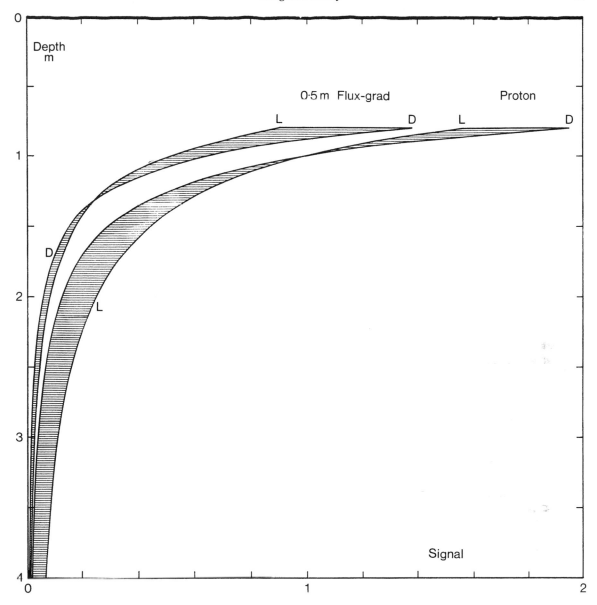

Fig. 59. Sensitivity decrease with depth for a total field (e.g. proton) instrument and a 0.5 m (1.6 ft) length fluxgate gradiometer, assuming an angle of dip of 70 degrees. L = linear feature; D = discrete feature.

that a ditch giving an 8 nT anomaly at 1 m would give 0.9 nT at 3 m (10 ft).

At depths shallower than 1 m (3.3 ft), sensitivity increases very rapidly. This can be used to estimate the depth of anomaly sources, as will be shown in the section on magnetic scanning below.

The standard height for carrying a detector of whatever kind is normally 20–30 cm (8–12 in.), which lifts it sufficiently to reduce the effect of 'noise' caused by variations in the magnetism of the topsoil, the depth of natural or contamination by iron or igneous stones, as well as problems caused by low vegetation. Sometimes, however, it can be beneficial to carry the instrument nearer the ground. In discussing above the effective depth of a ditch, we have assumed a triangular

Fig. 60. (Left) The ideal ditch: triangular section, with uniform topsoil filling (Weekley, Northants). (Right) An extreme example of a non-ideal ditch: this is mostly backfilled with non-magnetic chalk, and with only a lens of soil at the top (Badbury, Dorset).

cross-section of detectable soil completely filling it (Fig. 60 left). In fact it is much more common for the lower part of a ditch to be filled with primary silt mostly derived from the natural, while the more magnetic, usually darker, material is closer to the surface, sometimes forming hardly more than a lens at the top (Fig. 60 right). If a ditch with this type of filling lies on a lithology with low magnetic noise, such as the chalk of Fig.

Fig. 61. Idealized features showing positions of equivalent dipoles. (Left) Cylindrical (or square) pit. (Centre) Soil filled ditch like Fig. 60 left. (Right) Ditch with heavy primary silting, like Fig. 60 right. In this case, lowering the 0.5 m gradiometer from 30 cm to 10 cm (12 in. to 4 in.) will increase the signal by a factor of 2.2 for 30 cm (12 in.) topsoil depth, or 2.8 for 15 cm (6 in.).

60 (right), it is likely to produce a weak signal, and a worthwhile increase in sensitivity can be gained by lowering the magnetometer to about 10 cm (4 in.), as Fig. 61 demonstrates.

Heathcote (1983) has produced families of curves based on the theory of Linington (1973), characterizing the responses to archaeological features of a variety of shapes. This approach works well for a ditch with a uniform filling such as the one shown in Fig. 60 (left), but will be less helpful when there is a substantial proportion of primary silt or other non-uniformities.

Spatial resolution The high resolution continuously plotted fluxgate gradiometer survey shown in Fig. 56, and already described, is used to illustrate this discussion. The main traverse across the

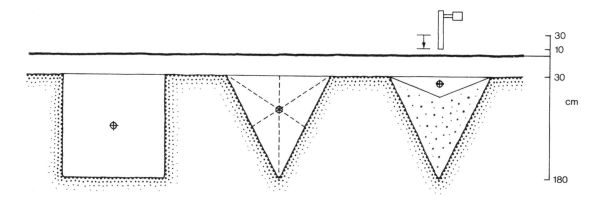

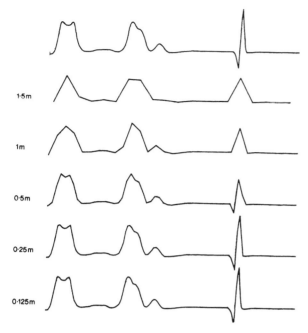

1·5 m

1 m

0·5 m

0·25 m

0·125 m

Fig. 62. Recording intervals and resolution. (Top) Continuously recorded trace, adapted from Fig. 55. The traces below show the resolution of this profile for the recording intervals shown.

kiln on the left – the most clearly resolved – has been transferred to Fig. 62 for analysis, with the addition of a typical positive and negative pair of spikes caused by near-surface modern iron rubbish. The kiln to the right does not give such a neat response and may be damaged, or of a different type. The weaker anomaly beside it on the right could be caused by a relatively narrow small kiln or pit. The traces below show how the response to these features would appear with readings recorded at different ground intervals.

An interval of 1.5 m is roughly equivalent to the 5 ft used in early work. The kilns are recorded strongly, but without any detail, and only one reading coincides with the iron spike so that it also looks like a kiln. Because of this sort of problem, a detailed search was needed around anomalies to get more information. The remaining four traces are at spacings often used for automatic recording. C is at 1 m (3.3 ft). The archaeological features are better resolved but the double peak of the iron spike is still not detected; it could be interpreted as a kiln, but perhaps a suspiciously narrow one.

With a 0.5 m (1.6 ft) recording interval, the detail of the kilns begins to resolve quite well and the iron spike is revealed by its sharp negative peak, although suppressed in height. The improvement compared with 1 m (3.3 ft) is very significant, and is clearly visible in practice. At a 0.25 m (10 in.) interval, all the detail of the kilns shows clearly. There is distortion in the steepest parts, with the iron spike rather broadened and slightly suppressed, but this would not affect interpretation. At eight readings per metre, a spacing of 12.5 cm (5 in.), the archaeological anomalies are as well resolved as in the direct record. The sharp rise and fall of the iron spike sets the most demanding test and, although its width is now correct, both the positive and negative peaks are slightly curtailed.

To summarize, a reading interval of 0.5 m (1.6 ft) is the largest suitable for detailed recording, and there is a further gain in resolution at 0.25 m (10 in.), four readings per metre. Going to 0.125 m (5 in.), eight readings per metre, produces only a marginal improvement on this. At four readings per metre, with 1 m (3.3 ft) between traverses, a 30 m (100 ft) square takes 3,600 readings, so that an FM18 magnetometer could record only one square at a time. An FM36, on the other hand, could record four squares.

There remains the problem, alluded to above in connection with direct plotting, of the poor resolution perpendicular to the traverses due to their relatively wide separation, normally 1 m (3.3 ft). It is perfectly possible to use narrower spacing. Spacing of 0.5 m (1.6 ft) can be achieved by making two surveys of the same area, staggered by 0.5 m (1.6 ft), then combining them at the plotting stage. This can be more satisfactory than actually trying to survey at the closer interval, because of the difficulty of modifying the standard 1 m (3.3 ft) routine. Often, in fact, not much seems to be gained by closer spacing nor by surveying an area twice in perpendicular directions. This is probably because the detailed measurement along the traverse lines is highly effective for characterizing the nature of the features, so that it is sufficient only to establish their extent in the other direction. An exception already mentioned is the case of linear features parallel with the traces. These may have no cross-traverses to characterize them and may displace only one traverse, sometimes in a confusing way. They are thus often difficult to see on a trace plot but will duly appear on dot density, grey-tone or even contour plots, though lacking in fine detail. The problem is relieved by chance which decrees that features truly parallel with the traces are quite rare, and that they are unlikely to remain parallel over the whole survey area.

Fig. 63. Calculated response of a total field (e.g. proton) instrument and 0.5 m (1.6 ft) fluxgate gradiometer to dipoles buried 0.5 m (1.6 ft) beneath the instrument at geomagnetic field inclinations of 0°, 30°, 60° (nearest to the British value of about 68°) and 90° (Heathcote, 1983). The dipole has a magnetic moment of 0.1 Am^2 and is parallel to the geomagnetic field, simulating a feature such as a high susceptibility small pit with induced magnetisation. Linear features give responses of similar shape, except that the negative peak is stronger when they run approximately east–west (Linington, 1973).

Magnetic anomalies: shapes and forms

It may appear that this fundamental subject should have been discussed earlier, but it seemed worthwhile to start with a few straightforward practical examples to give a general idea of what we are looking for, before going into the subject in detail.

Fig. 50 shows in diagrammatic form a mass of baked clay representing a kiln. The parallel lines are the Earth's magnetic field, and the lines of force induced in the kiln by this field are curved. As with any bar magnet, the 'magnetic circuit' is completed by the external return flux, which is of opposite polarity to the magnet. Because of this return flux, every positive magnetic anomaly is accompanied by a lesser negative anomaly alongside it; and because of the angle of dip (or inclination) of the field in northern latitudes, more of

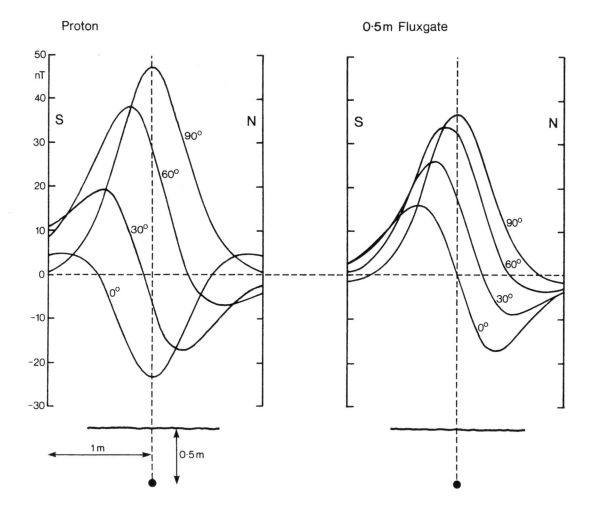

the return flux is exposed on the north side of a feature than the south, and the negative anomaly is therefore bigger on the north side. Another effect of the angle of dip is to displace the maximum of the anomaly slightly to the south. Negative anomalies can be seen in Fig. 55 as downward displacement of some of the traverses. This pattern is reversed in similar latitudes of the southern hemisphere.

Moving from the North magnetic Pole to the Equator, changes occur in the shape of anomalies as the field direction ranges from vertical at the pole to horizontal at the magnetic Equator. In Fig. 63 the effect of these changes on a total field instrument, such as a proton magnetometer, is compared with their effect on a fluxgate gradiometer. In the first case, the anomaly is actually reversed at the two extremes, with substantial changes of form in between. The responses to iron-working hearths in Nigeria, close to the Equator (Fig. 64), are in agreement with the theory (Tite, 1966).

Fig. 64. A survey of iron-working furnaces in Nigeria by proton magnetometer, a practical demonstration of the negative anomaly effect close to the Equator (Tite, 1966).

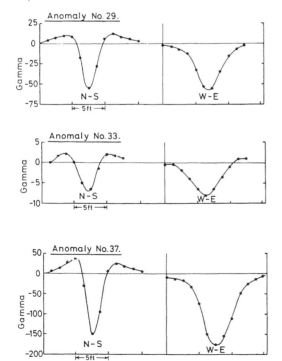

The fluxgate gradiometer differs in measuring only the gradient of the vertical component of the field, and in being sensitive to its direction. For these reasons, its response patterns and their changes with latitude are very different from those of the total field instrument as one approaches the Equator, but less dramatically variable in that the positive component of the anomaly always remains significant. Thus the survey of the wadi-bed site of Al-Ghubra in Oman (Fig. 65), which, at 22.5°N of the Equator, is only about 10° further north in latitude than the Nigerian survey, gives responses that look quite normal to eyes accustomed to British surveys. With a proton magnetometer, the negative anomalies would be dominant.

A useful rough rule for practical interpretation is that the width of the anomaly at half its maximum reading is equal to the width of the buried feature, or its depth if this is greater. The use of this rule in scanning is mentioned below.

Magnetic scanning

Magnetic scanning is another important development in archaeological prospecting made practicable largely by the fluxgate gradiometer. So far, the description of magnetic survey techniques has assumed intensive surveys designed to extract the maximum information from a site for subsurface planning. The very word 'prospecting', though, suggests wide-ranging surveys spread over the countryside in search of the unknown. It became clear early on that, providing the local lithology was suitable for producing clear magnetic contrasts, the fluxgate gradiometer was ideal for this, without direct recording but relying on the operator to recognize the response of the instrument to archaeological anomalies. Its use in this way also became urgent: large tracts of land subject to development needed to be checked for their archaeological potential and the areas involved made conventional detailed survey quite unthinkable. Scanning can be a valuable complement to air survey and is effective where there is woodland, permanent pasture or other ground cover impenetrable to the aerial view.

In scanning, speed and broad assessment are the keynotes. The operator needs to be experienced in anomaly interpretation, especially of trace plots, which he needs to see in his imagination as he looks at the instrument dial – an

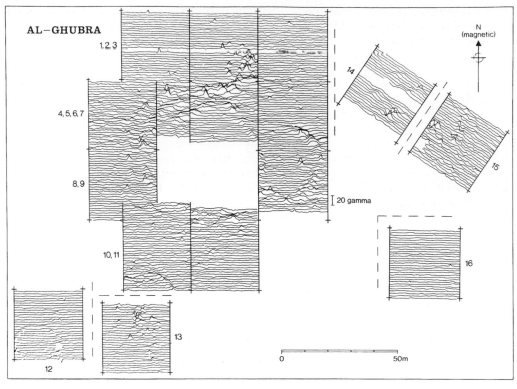

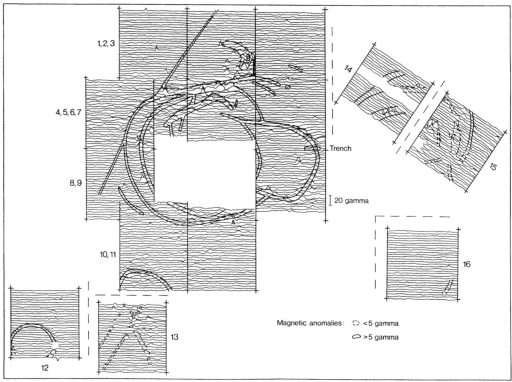

Fig. 67. Magnetic scanning.

Fig. 66. Sectioning one of the Al-Ghubra ditches (trench shown in Fig. 65), which proved to be almost 3 m (10 ft) deep.

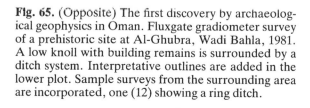

Fig. 65. (Opposite) The first discovery by archaeological geophysics in Oman. Fluxgate gradiometer survey of a prehistoric site at Al-Ghubra, Wadi Bahla, 1981. A low knoll with building remains is surrounded by a ditch system. Interpretative outlines are added in the lower plot. Sample surveys from the surrounding area are incorporated, one (12) showing a ring ditch.

analogue, or pseudo-analogue output is essential, as digital numbers do not enable one to picture the survey in this way.

Bamboo flags are used to mark promising anomaly positions. These need to be quite long, about 4 ft (1.2 m), if they are to be used in heavily vegetated areas, otherwise 1 m (3.3 ft) is a good length. They can be marked with either flags made from wide, red adhesive tape or upholstery tape, or fluorescent labels from stationers. An auger is a valuable ancillary tool for testing anomalies without distinctive archaeological shapes. A hand-operated coring type, up to 30 mm (1.2 in.) in diameter and 1 m (3 ft) long, is quick and effective. Needless to say, its use must be restricted to sites where it is not likely to do damage, as it can to a Roman mosaic or assemblages of pottery. Magnetic scanning can be divided into five fairly distinct types of application:

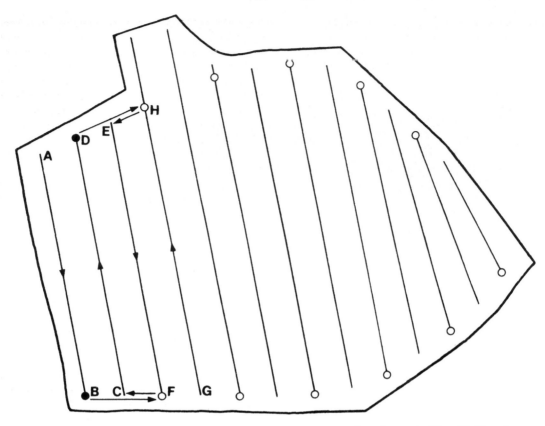

Fig. 68. Intensive scanning method. Notional baselines follow the top and bottom of the field. A ranging pole is put in at B. Assuming a two-pace (approximately 2 m or 6 ft) traverse spacing, a ranging pole is put in at D, two paces from A. The first traverse is from A to B, and B is then moved four paces to F. The second traverse starts at C, two paces back from F. When D is reached, it is moved four paces on to H, and the third traverse starts two paces back at E, and so on. The two base-lines can follow quite erratic field boundaries – the odd shape of the field has been chosen to demonstrate this. The effect of following an angular course is simply to make the ground coverage more intensive. The projection at the top is scanned by extrapolation beyond the ranging poles.

1. Detailed scans Scanning was first tried for fairly intensive examination of limited areas, usually one field at a time. A traverse spacing of 2 m (6 ft) gives a good compromise between speed and coverage, ensuring that reasonably substantial pits, hearths and kilns will be detected, as well as linear features. The survey is pursued by walking towards ranging poles set up along the two longer sides of a field or similar area, moved as shown in Fig. 68. Bamboo markers are pushed in where promising anomalies occur, and if these do not form an obviously man-made pattern, more intensive tests are made around them at the end of the main scan.

2. Hillforts A broader version of the type 1 scan has proved highly effective for rapidly assessing the density of occupation of Iron Age hillforts in England. It is capable of distinguishing those that were continuously occupied 'towns' from those built only as refuges, containing only slight signs of occupation, perhaps during the building operation, or possibly a small group of storage pits for emergency supplies. The following abstract from my description of the work on three Surrey hill-forts, all on Greensand and heavily wooded, seems a suitable introduction to the approach, and some of its problems, physical and psycho-logical:

> ... By measuring the heating effect in labora-tory conditions, Tite has shown that the [mag-netic] enhancement in Lower Greensand soils is particularly strong (Tite, 1972b). Suscepti-

bility measurements ... on samples from Holmbury gave values of 25×10^{-8} SI/kg for bedrock, and 59×10^{-8} SI/kg for topsoil. With these values, a topsoil-filled pit 1 m [3.3 ft] in diameter and depth, with its top 0.3 m [1 ft] below the surface, would give a detectable magnetic anomaly of 2.4 nanotesla [adjusted for low values given by the susceptibility bridge used at the time] with a 1 m [3.3 ft] fluxgate gradiometer held 0.3 m [1 ft] above the ground; and, after enhancement by human activity, much smaller features would be detectable. The thermoremanent magnetism of hearths on the Greensand could also be expected to be high. It therefore seemed likely that magnetic surveys of the three hillforts ... could produce a comprehensive indication of the extent of their occupation before excavation.

A normal intensive survey using a grid was precluded by density of vegetation and lack of time, so each hillfort was scanned with ... the gradiometer, the readings of which are continuous and were directly interpreted, and recorded only when archaeological anomalies were recognized. The basic scan consisted, as far as trees and undergrowth allowed, of parallel traverses about 10 m [30 ft] apart, filled in more closely where indications of occupation were detected. Additional scans, including a perambulation just behind the inner rampart, were also made in sheltered parts of the fort where occupation seemed especially likely to have existed. Magnetic anomalies were few enough to be located individually by measuring their distances from suitably marked adjacent trees. The nature of some of the anomaly sources was checked with a 1-inch [30 mm] coring auger ... typical magnetic gradients for both pits and hearths at Holmbury and Hascombe were 6-13 nT.... Each survey was completed in less than a day.

... the almost total lack of archaeological anomalies in the first survey, of Anstiebury, came as such a surprise that it was assumed that magnetic contrasts were unaccountably suppressed on the site, so that unduly small anomalies, in fact caused by bedrock irregularities – or 'soil noise' – were identified for excavation. This experience, and the presence of features giving normal anomalies at Holmbury and Hascombe, enabled the surveys of these sites to be more rationally interpreted. (Thompson, 1979.)

The hearths at Holmbury proved to be almost the only internal features present, and could well have been the campfires of those who constructed the defences. At Hascombe, the scan detected a group of pits containing very early coins, and burnt wood and grain. This made possible an extensive exercise in radiocarbon dating and a very useful calibration point for the archaeomagnetic dating curve, which was being developed at the time, and a date for activity in the hillfort between 70 and 50 cal BC. As the last paragraph shows, subjectivity in interpretation is a danger against which one must always be on self-critical guard.

There has been little demand for the magnetic scanning of hillforts since this work was done, but with minimal resources it could provide a comprehensive picture of which hillforts were centres of population, possibly tribal capitals, and which were simply built as refuges and in some cases never used. Obviously this kind of survey is not confined to hillforts, and is applicable to large areas generally.

3. Motorway routes Although just another example of the large area problem, motorway routes have benefited from a specially adapted approach that is worth describing. The M3 route through Hampshire was the first major operation of this kind, and again it seems appropriate to quote a passage written for the report:

The M3 project was the first to which magnetic scanning was applied on a landscape scale. Its use was necessary to help to ensure that no significant archaeological sites were missed in the broad strip of country, several miles long, that was to be affected by the road. It was specially effective as a complement to air photography and fieldwalking in areas of woodland and permanent pasture. For such a survey to be achievable, there must be a clear contrast in magnetic susceptibility between subsoil or bedrock and topsoil, so that silted archaeological features are readily detectable. The chalk in central Hampshire, overlain by topsoil enhanced magnetically by an element of clay-with-flints, provided these conditions, although natural pockets and spreads of the clay-with-flints sometimes caused misleading indications.

The process used on the M3 was extremely simple. The survey was tackled field by field by two people, one operating the instrument, the other doing auger tests and plotting. Three longitudinal scans were made, two just within the sides of the potential route, and the third along the centre line, and finally two diagonal lines to provide an extra check and slightly closer coverage. It seemed reasonable to assume that no archaeological site of any substance, especially those including ditches, would be missed by this density of coverage of a swathe of countryside averaging about 60 m [200 ft] in width. Magnetic anomalies of archaeological type were marked with flagged bamboos. The survey was intensified around these in order to define the plan of the site, and further flags inserted. If these formed linear or circular patterns, a human origin was assumed, but the positions of discrete anomalies were examined with a coring auger, as they were sometimes due to natural pockets of clay-with-flints. It was possible to cover about a mile [1.6 km] a day, including the plotting of features discovered. In one case (London Lodge), the scan was followed by an intensive recorded survey because of the difficulty of distinguishing features of some complexity and subtlety in the presence of magnetic 'noise' due to clay-with-flints.

The scanning scheme used is illustrated in Fig. 69. It merits some explanation, because purists have complained that it gives uneven ground coverage, and that one should zigzag across the road strip in the manner of the intensive scan. The method was adopted as the most practicable for the very early stage at which the M3 route was scanned. Nothing was marked on the ground, and the route had to be established by using the

Fig. 69. Road route scan (see text).

road plans to measure in the future position of the road along hedgerows. Four ranging poles placed at these points were usually the only guidance for the scan, so that longitudinal traverses between them were the most feasible. Also, with minimal changes of direction, any problems of directional sensitivity of the instrument were minimized. Up to about 2.5 km (1.5 miles) a day can be covered if little is found, so that a 20 km (12.5 mile) stretch could be scanned in just over eight days. If the work had been taking place after the installation of fences, the zigzag scheme would have been more attractive, especially with more modern and stable instruments. A positive advantage of the long traverses with modern equipment would be their use in combination with a recording device of the ST1 type (see below), because they can be readily relocated by means of the four ranging poles.

The part of the M3 route investigated was about 20 km (12.5 miles) long. The following discoveries were made: two groups of ditches and pits (including London Lodge, above); an Iron Age 'banjo' enclosure in woodland; a ring ditch with an associated linear ditch; a pair of linear ditches, probably part of a drove road previously seen on air photographs some distance away; and a ditch probably belonging to an extension of a known Iron Age site. In addition, an oval mound found in fieldwalking was shown to be surrounded by a ditch, and therefore a barrow, and a central cremation area was also detected. Misleading results were produced in some cases: a barrow seen on air photographs contained strong anomalies interpreted as possible Saxon secondary burials with iron weapons, which in fact proved to be modern wartime rubbish, and an important area of occupation with enclosure ditches was not detected in the scan, although visible on

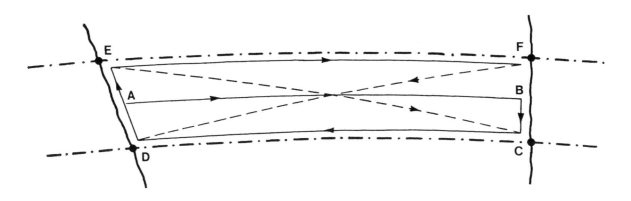

air photographs. The ditches were quite slight and contained much chalk, so that they gave little magnetic contrast, which was swamped by the 'noise' generated by plough furrows.

A few scans have since been made on other road routes, including the Bridge by-pass in Kent, where a Saxon cemetry was discovered, and a road widening scheme at Baylham House (*Combretovium*) in Suffolk, where Romano-British pottery kilns and the side ditches of a Roman road were found.

Scanning has been found particularly effective for tracing 'ribbon development' along Roman roads, which is often delimited by a clearly detectable ditch parallel to the road along the rear of the properties. Later ridge and furrow can be peculiarly helpful in enabling one to use a magnetometer for defining broad spreads of occupation-enhanced soil to which it is normally insensitive: the concomitant variations in soil depth cause distortions in the Earth's field that the magnetometer detects increasingly clearly as the focus of ancient occupation is approached. This is really the plough furrow noise effect extended to a longer and regular wavelength.

4. Fieldwalking The use of a fluxgate gradiometer as an adjunct to standard fieldwalking techniques is much to be recommended. To the archaeologist it is (or could be, if more exploited) an invaluable 'third eye', not only usable in its own right but enabling him to check any site suspected from artefact scatters: not only can it establish whether actual features exist on the site, but the size of anomalies will give some indication of their state of preservation. Ideally, magnetic susceptibility measurement should also be used as it is sensitive to sites which are more superficial.

5. Preliminary testing This has already been alluded to. Scanning is a necessary preliminary to detailed magnetic surveys in that it provides information on the size of signals likely to be produced by the features, and can also give an indication of the extent of remains so that the detailed survey can be economically planned. When considering a site for future survey, as we shall see in the next chapter, sensitivity of response can be estimated by making magnetic susceptibility measurements on the topsoil and natural, with the advantage that samples can be sent by post for assessment before the site is even visited by a geophysical team. On site, broadly spaced susceptibility field measurements can give a useful indication of areas worth surveying, especially on sites too large for complete magnetometer survey where preliminary information is lacking. But a magnetic scan, except where soil noise is so high or modern rubbish so dense as to give misleading readings, is also valuable because it gives an indication of how the survey will actually look.

Scan evaluation

As stated above, the standard approach when promising anomalies are encountered in a scan is to push in bamboo markers. These may well build up into a recognizably man-made pattern as the survey proceeds. Otherwise it will be necessary to check out such findings, either by an immediate local intensification of the scan or by returning later. If no clear pattern emerges, it may be necessary to resort to cautious use of the auger.

Weak anomalies may be caused by features to the side of the scanning line; this can be checked by moving sideways from the position of the peak value. A valuable trick for rapidly checking the credibility of an anomaly, especially whether it is caused by a piece of iron rubbish, is to stand at its centre and lower the instrument from a height of 30 cm (1 ft) to the ground. The shallower the feature, the bigger the change in reading, as shown in Fig. 71. A piece of iron 20 cm (8 in.) deep will give a reading change of very roughly 17 times, and anything shallower a great deal more. The very shallowest archaeological feature, centred at perhaps 30 cm (1 ft) deep, would give a change of nine times, so a factor of 10 might be regarded as the cut-off level if the topsoil is not more than 20 cm (8 in.) deep. At 1 m (3.3 ft) depth the factor is 2.5, at 2m (6.6 ft) it is 1.7 and at 3 m (10 ft) it is 1.5. In other words, at most 'archaeological' depths the factor hovers around 2.

Another way of recognizing iron rubbish is to look for sharp, randomly placed negative peaks; but sometimes the negative pole is quite deep and not readily seen, and deep iron generally is more difficult to distinguish. Some soft iron has negligible permanent magnetization and acts as a high susceptibility object, giving an anomaly like a miniature pit. Even with normal iron, the 'two-

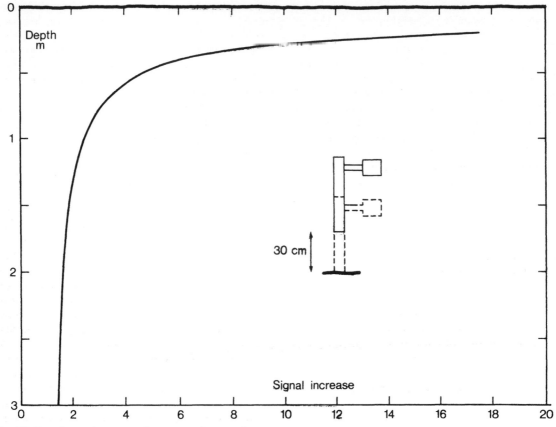

Fig. 70. Depth test for anomaly sources (see text).

Table 3. Magnetic anomalies produced by small pits of two sizes at various depths, with a susceptibility difference of 100 SI units between the filling and the surrounding subsoil. The detector is assumed to be 30 cm (12 in.) above the ground. The values are for an absolute instrument, and the reduced signals from a 0.5 m (1.6 ft) fluxgate gradiometer may be obtained from Fig. 59 – remember to use the depth of the centre of the pit. Anomaly strength varies in proportion to magnetic susceptibility.

Susceptibility difference 100×10^{-8} SI/kg

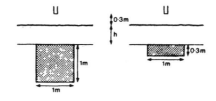

h (metres)	Magnetic anomaly (nT)	
0.3	4.1	2.3
0.6	2.1	1.1
0.9	1.2	0.6
1.2	0.7	0.3

Fig. 71. Surveying the Ring of Brodgar, close to Stenness, with the 1 m (3.3 ft) Plessey fluxgate gradiometer.

height test' can be quicker to use than to scan around the anomaly position.

I have already mentioned the approximate rule that the width of the response peak at half the maximum signal is equal to the width of the buried feature, or its depth if this is greater. Thus in scanning, one notes the maximum signal and then the outline of the feature can be determined by marking points where the reading is half the maximum – having ascertained that it is not too deep by using Fig. 71.

Given some investment, the state of the art is now such that the effectiveness of magnetic scanning could be considerably enhanced. The ST1 survey recording system could be used to picture the traverses on a computer screen so that they could be analysed in as much detail as those of a

formal survey, saving much subsidiary scanning and the need to make assessments in real time on the move. This will speed up the actual scanning process and improve its reliability, but, as markers will be put in only after the traces have been examined, relocation of anomaly positions will depend either upon well marked traverse lines or a quick additional 'old-fashioned' scan. At present plotting in discoveries is often done with a plane table, or simple tape and compass readings, which could be replaced by an electronic distance meter (EDM).

It must be added that the methods of detailed recording are now so fast that there is a tendency, at least for smaller areas, to replace scanning with full surveys (for instance, sample areas in hillforts) as well as to add detail where positive results are obtained.

There will be some further discussion of scanning in Chapter 6.

Geology and magnetometer surveys

Sedimentary areas We have seen that the effectiveness of magnetometer detection depends
upon the absolute magnetic susceptibility of the
soil, and its contrast with that of the underlying
subsoil. Tite (1972b) measured more than 100
samples of ordinary agricultural topsoil from the
southern and eastern parts of England and
related these to a geological map to give an indication of the likely success of surveys in different
areas. He prepared the samples by heating them
in a reducing atmosphere of hydrogen to bring
about the maximum conversion of the iron compounds present to a strongly ferrimagnetic form,
thus providing a figure for the concentration of
iron oxide in the soil available for conversion.
The percentage of the original susceptibility
value to the value after this treatment ($100 \times \chi_O/\chi_H$) gives a figure for the fractional conversion
of the iron oxide which has occurred *in situ*; but
more important is the difference between the
values, which gives an indication of the increase
in magnetic contrast that could develop on an
archaeological site. More realistic conditions
seem to be provided by an alternative treatment,
which was also used, of heating in nitrogen followed by air, in which the reducing atmosphere is
provided by the organic content of the sample it-
self·(Tite and Mullins, 1971).

Tite tabulates his results in detail. Here they
are summarized and augmented from measurements collected by the Ancient Monuments Laboratory, with the addition of some data from
overseas (Tite and Linington, 1975).

The Jurassic Ridge, which includes the Cotswolds, and runs like a backbone across England
from Dorset to Cleveland, provides some of the
best surveying conditions, the iron-rich topsoil
contrasting strongly with the underlying limestone. Drift deposits derived from this, including
the silts and gravels of rivers which rise on the
ridge, such as the Thames, Nene and Welland,
are also good. Other responsive sedimentary
strata are the Lower Greensand, and the Carboniferous and Devonian of the south-west peninsula, with the exception of the uncultivated
moorland soils on granite, such as Dartmoor and
Bodmin Moor. The Ordovician areas of Wales
seem to provide good conditions, as far as they
have been tested. Cretaceous deposits such as the
Upper Chalk of Salisbury Plain are generally

weak, but stronger where they are overlain by
such Tertiary material as clay-with flints (see
Chapter 4) or are close to such areas, as in much
of Hampshire and on and around the Chilterns.
We have also seen that there is a boulder clay element in the soil on the chalk of the Yorkshire
Wolds which can enhance response there.

Much of East Anglia is covered by weakly magnetic drift deposits, or is affected by periglacial
phenomena or glacially transported material,
which can be confusing. Thus archaeological
prospecting of all kinds can be difficult in the
area. Across the country, in the region of the
Welsh Marches, the extensive areas of Keuper
Marl and associated deposits are also rather
unresponsive.

The gravel terraces of some rivers present difficult conditions for magnetic survey, natural
channelling effects often dominating the archaeological. It seems that iron concretions in the gravel seem to be largely responsible, producing
their own magnetic patterns and reducing the susceptibility contrast with the topsoil.

Studies on soils from abroad have focused on
Italy, where soils of the *terra rossa* type, mostly
developed on hard limestones, showed significantly higher natural percentage conversions
than most British soils. Similarly enhanced values
were found on limestones and calcareous alluvia
from Greece, Crete and Turkey, and Tite and
Linington surmise that this is probably due to the
alternation of damp winters with hot, dry sum-
mers encouraging the fermentation effect discussed here in Chapter 4. In the Tropics, similarly
strong enhancement occurs where there is an
alternation of dry and wet periods, but not in regions where the climate is continuously humid.

Table 3, adapted and simplified from one
published by Tite (1972b), provides a 'ready
reckoner' for detectability to be used alongside
site susceptibility contrast figures. The anomalies
vary linearly with the susceptibility difference.

Igneous areas Much of the Highland Zone of
Great Britain is composed of igneous rocks with a
substantial content of iron compounds which acquired a thermoremanent magnetization when
the rock first cooled. In areas of the kind
favoured by ancient man, such rocks rarely form
a continuous subterranean sheet, but are broken
up into individual boulders or pebbles deposited
by glacial and water action in valley bottoms, or,

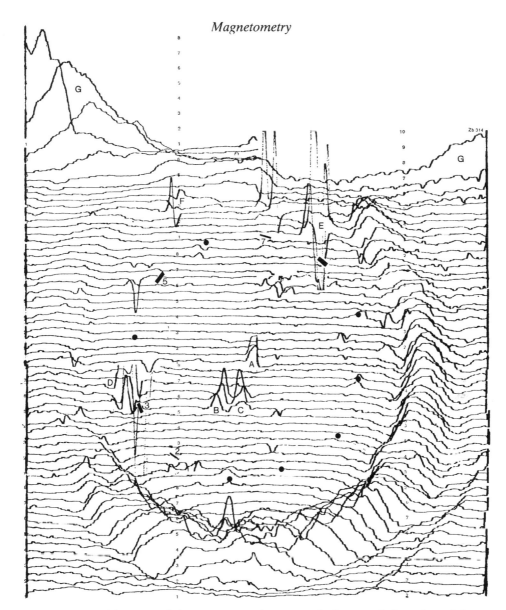

Fig. 72. Standing Stones of Stenness, Orkney: survey with the 1 m (3.3 ft) fluxgate gradiometer, width 60 m (195 ft). The obliterated parts of the surrounding ditch of this henge monument show as a magnificent anomaly reaching 50 nT, with an entrance on the north. Extant parts of the ditch are weaker, as they contain less magnetic soil. A, B, C: ritual pits; D, E, F: iron spikes. G: natural anomalies caused by intrusions of igneous rock, fortunately just clear of the monument. (Note the lack of negative anomalies associated with this deep-seated source – compare Fig. 75.) 2, 3, 5, 7: stones still standing. Solid circles: positions of eight stone holes found in excavation; five were detected in the survey, allowing the others to be predicted by interpolation, all to within 1 m (3 ft). Sensitivity: 15 nT/line interval.

as in the Orkney Isles, form narrow dykes thrust up through fissures in less magnetic sedimentary rock. Igneous rock is also found as glacial erratic material well south of the Highland Zone. When they have been displaced, the magnetic directions of the rocks are randomly jumbled, producing 'noisy' background signals that normally dominate the effects of archaeological interest.

The St Lucia magnetometer survey (Chapter 4) is an example of the noise effect produced by loose pebbles and boulders. The site of the Standing Stones of Stenness in Orkney (Ritchie, 1978), where the archaeology is so dramatically defined, is just missed by a massive basalt dyke which could have completely vitiated the survey if a few metres further south (Fig. 72). A survey at Skara

Fig. 73. The east terminal of the Stenness ditch. Note the high magnetic susceptibility boulder clay overlying the low susceptibility Old Red Sandstone bedrock and filling the ditch, giving strong anomalies.

Fig. 74. Stenness: the central ritual pit, A, with stone surround. This contained burnt bone, and the anomaly is caused by magnetic enhancement of the upper layers by heat.

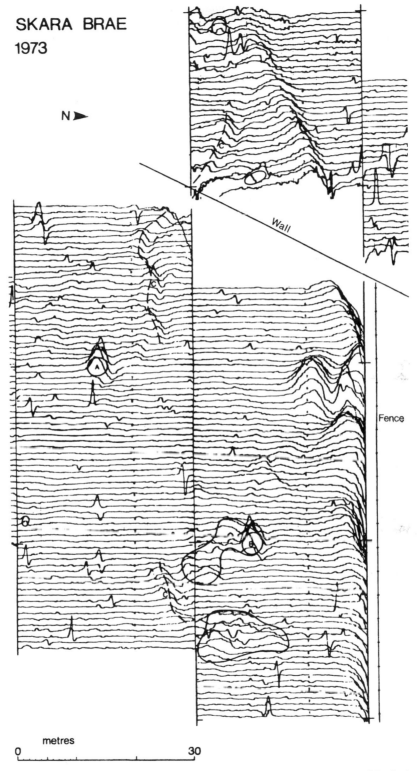

SKARA BRAE
1973

N ▶

Wall

Fence

metres
0   30

Fig. 75. The Skara Brae survey, showing the linear natural anomaly running from top to bottom. Anomalies of archaeological type, with distinctive negative lobes, are outlined or marked with broken lines.

Brae (Fig. 75), to ascertain whether there was any extension inland of the famous Neolithic village, produced a linear feature at first suspected of being a ditch, but it lacks the clear attendant negative anomalies associated with the more archaeological-looking anomalies, for instance A and C. This characterizes it as much more deep-seated, with return flux much more spread out than that of compact features – in effect a monopole, with its opposite pole deep underground. There is also curious behaviour at the centre of the survey, where the smoothness of its line is interrupted and the main anomaly is displaced southward – again more natural in appearance than archaeological. The interpretation of this feature was not helped by the coincidence of the modern fence along much of its length, which had the effect of superimposing a negative anomaly just where it would have been if the feature were indeed archaeological.

A completely different state of affairs obtained on the island of Iona, where a survey was made in an attempt to trace the full length of the Dark

Fig. 76. Orkney: an igneous dyke exposed on the foreshore.

Age ditched boundary known as St Columba's Vallum, around the Abbey complex (RCAHM, 1982) The dramatic basalt pillars of Statta are visible from the site and igneous rocks abound, so that it seemed hardly worthwhile even to take a magnetometer to it. In fact, tests over a known ploughed-out part of the Vallum ditch produced a clear negative anomaly. A small area survey was made (Fig. 76) at a critical point where the Vallum, coming down towards the sea, had previously been supposed to change direction towards the Abbey, following a visible bank. This survey showed the ditch continuing on its straight course, while the bank was represented by a band of 'noisy' traces with no accompanying ditch anomaly.

In previous surveys, ditches had always been represented by positive magnetic anomalies. The Abbey complex of Iona is situated on some of the rare fairly level ground on this craggy island. The ground is a raised beach, which must contain, in this area, a substantial proportion of magnetic grains which would have tended to align with the Earth's field during settlement in water, leaving the beach with an overall detrital remanent magnetization – the phenomenon which has enabled us to build up a history of the behaviour of the geomagnetic field from the sediments of lakes (Clark *et al*, 1988). The digging of a ditch, followed by its backfilling by weathering and ploughing, would have created a band of disordered particles with no net magnetization, or far less, which would therefore have been detectable as a negative anomaly. The noisy traces along the line of the bank showed that this was merely a field bank, along which magnetic stones cleared from the field had been deposited in the customary way. Subsequently, palaeomagnetic measurements on some samples from undisturbed subsoil on the site confirmed that they did retain a weak, fairly consistent remanent magnetization.

The knowledge gained from this first, simple survey was invaluable when the Iona survey was continued in the more complicated area to the south of the Abbey, where the course of the Vallum was completely unknown. At least two phases of intersecting ditches were revealed as negative anomalies, and later confirmed by excavation. One seemed to mark an extension of the Graveyard of the Kings. There was also evidence that some lengths of ditches may have been back-

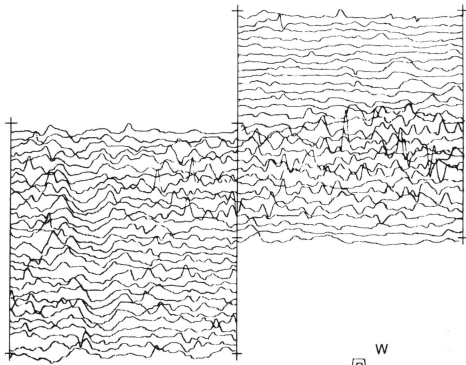

Fig. 77. St Columba's Vallum, Iona. Two 30 m (100 ft) squares. Sensitivity: 7.5 nT/line interval.

filled at least partly with stones, so that their courses were marked by lines of erratic readings rather than negative anomalies. Close to the monastic buildings, more familiar positive anomalies appeared to mark the line of a ditch, perhaps because here the remnants of fires and other domestic detritus may have produced a magnetic enhancement of the backfilling soil which dominated the subtractive effect seen elsewhere.

A survey of a small ditched enclosure at Shiels, on the gravels of the River Clyde just downstream from Glasgow, produced comparable but subtly different results. Here the background was quite noisy, presumably due to fairly large magnetic stones in the gravel. Rather than as negative anomalies, the ditch was represented by just-recognizable lengths of relatively smooth trace, which must have been due to a silty filling without stones. This site, like Neatham (Fig. 55), would

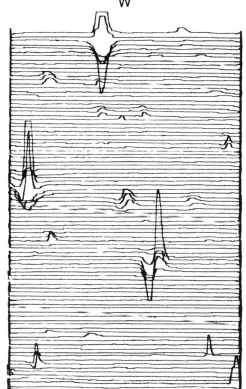

Fig. 78. Magnetic survey of a Saxon cemetery, Pewsey, Wiltshire. Strip 30 m (100 ft) wide; traverse interval 0.5 m (1.6 ft). Sensitivity 7.5 nT/line interval.

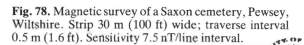

have been undetectable without the detail present in the traces.

The magnetometer as a metal detector

Very occasionally, the high sensitivity of magnetometers to iron is made use of in archaeology. Fig. 78 shows a magnetic survey of a Saxon cemetery at Pewsey, Wiltshire. At least some of the graves produced strong, sharp anomalies, with negative peaks tending to show as strongly as positive. These are the graves of warriors with their swords lying horizontally at their sides. Other graves have lesser anomalies, perhaps due to such accoutrements as knives and spearheads, while some are without iron and undetectable. Such cemeteries may be reasonably defined as long as a proportion of the burials contain iron goods.

It is interesting that the anomalies tend to run east-west, the direction of the survey, especially two of the three strongest in which the positive and negative anomalies directly overlie each other on the chart – clear evidence of the orientation of the graves. The 0.5 m (1.6 ft) traverse interval, made by superimposing two staggered surveys, was worthwhile for adding conviction to the smaller anomalies.

Chapter 4
Magnetic susceptibility

The study and utilization of the magnetic susceptibility of soils are part of the developing subject of environmental magnetism, perhaps the most sensitive and widely applicable parameter in landscape studies (Thompson and Oldfield, 1986). Its importance in magnetometer surveying has already been discussed, but it is now being increasingly exploited for archaeology in ways beyond the reach of magnetometers.

The usefulness of magnetic susceptibility in archaeological prospecting derives from two factors: the normally greater susceptibility of topsoil compared with the underlying layers, and the enhancement of this susceptibility by the activities of human occupation. In prospecting, features silted up with such soil are the province of magnetometer survey, while susceptibility prospecting has been particularly developed for detecting evidence of occupation, and defining its limits, in the topsoil layer itself: it does not require the existence of distinctive features. Thus sites which may only have survived in the topsoil can be detected, sometimes in quite fine detail. This topsoil work also makes a virtue of the fact that susceptibility instruments tend to have poor depth penetration compared with magnetometers.

Another major use of magnetic susceptibility occurs in the completely contrasting situation where topsoil is no longer *in situ*, but has been eroded and re-deposited in locations such as lake basins and estuaries. The displaced soil can be identified in cores and sections even if the ancient landscape to which it belonged has been swept from the face of the Earth. This is particularly valuable in areas where archaeology is elusive, and where it is uncertain whether human occupation has existed at all in the past. The stratigraphy of such deposits also provides a sequential picture of landscape evolution in which magnetic susceptibility peaks can reveal phases of clearance and exploitation.

Definitions

Because of the magnetic fields generated by their orbiting electrons, all atoms react to magnetic fields, and have a magnetic susceptibility, which is denoted by the Greek letter kappa (κ). This is commonly slightly negative, when it is known as diamagnetism. Materials with positive susceptibility become magnetized by a magnetic field, an effect especially strong in iron and its compounds. Unlike permanent magnetism, it is only measurable in the presence of the magnetizing field, and is defined as the ratio of the intensity of the induced field to that of the magnetizing field, or $\kappa = M/H$. Because M and H are measured in the same units in the SI system, these cancel out and the ratio is basically 'dimensionless'. Therefore, it does not need any unit to define it, although 'SI' needs to be added to distinguish readings from those in the old cgs system, which are numerically different.

In practice a large sample will give a stronger signal in a measuring instrument than a small one. Readings can be related to a standard size of sample, 'volume-specific'. Although commonly used, this can be imprecise because one often does not know how closely packed the material is – how much of the volume is the material being measured, and how much is air, water or pebbles. The best results are obtained by drying and sieving to remove these biasing effects, and then weighing the sample to produce 'mass-specific' readings which give better standardization. All this applies to laboratory measurement on prepared samples: field measurements are more rough and ready, and can only be calibrated in terms of volume susceptibility.

The symbol for mass-specific measurements is the Greek chi (χ) with the relationship $\chi = \kappa/\rho$, where ρ is the density of the material (not to be confused with resistivity). The introduction of the extra variable of mass means that the readings are no longer dimensionless, and are expressed as SI/kg. κ and χ are often called K and X in everyday parlance (see SI in Glossary).

Various kinds of magnetization are mentioned below. They are designated according to types of crystal structures, which define the extent to which the magnetic effects of the electrons in a crystal lattice reinforce or balance out one another. These are well illustrated in Thompson and Oldfield (1986). In *ferromagnetic* material such as iron, there is maximum reinforcement, and magnetization once acquired is permanent, while in *antiferromagnetic* structures there is complete balancing out. There are various intermediate states. In *canted antiferromagnetic* materials, there is a small residual susceptibility due to slight misalignment of antiparallel magnetic moments, a property shown by the oxide haematite (αFe_2O_3). *Ferrimagnetism*, exhibited by magnetite (Fe_3O_4) and maghaemite (γFe_2O_3), is similar to ferromagnetism but diminished by one in three of the moments due to iron atoms being in opposition to the remainder. These last two conditions carry some permanent magnetization, but this is greatly increased by the presence of an external magnetizing field. Another relevant magnetic property is *paramagnetism*, displayed by materials whose atoms or molecules have net magnetic moments which are aligned in the presence of an applied field. Paramagnetism disappears if the field is removed.

The magnetism of soils

We must now look more closely at the archaeologically valuable magnetic properties mentioned above, whilst bearing in mind that they are infinitely variable and imperfectly understood when applied to a complex material such as soil.

The mineral matrix of soil is mainly derived from natural bedrock by weathering, root action and human activities such as ploughing; its magnetic properties are largely determined by the iron content of this parent material. Iron compounds are relatively insoluble, and therefore tend to concentrate in the soil as the formation process continues. There are also other inputs, which can be illustrated by the example of chalk, which, although seemingly the archetypal nonmagnetic rock, can support quite respectably magnetic soils. One component is a mixture of clay (mainly the complex mineral montmorillonite) and iron oxides, distributed through the chalk and derived by wind or water action from land surrounding the Cretaceous seas. The oxides are concentrated by the substantial subaerial denudation of other components which has occurred over millions of years: we are familiar with the fact that natural chalk under Bronze Age barrows, or the bank of Stonehenge for example, is higher than the surrounding chalk which has not been similarly protected – and the Bronze Age is but yesterday in the history of soil formation. Further inputs to chalk soil derive from the detritus of Pleistocene glaciation, and earlier overlying Tertiary sediments. These, often modified by later processes, are represented by deposits broadly known as clay-with-flints, which still overlie the chalk in many places and in others have eroded away but left behind a substantial magnetic component in the chalk soil (Limbrey, 1975). Such processes have also contributed to the magnetic content of soils on other lithologies, with the added ingredient in areas close to the Highland Zone of glacier-transported igneous material.

Thus, there is a natural tendency for topsoil to accumulate iron minerals. Most common in temperate and humid climates is the hydroxide goethite ($\alpha FeOOH$), while haematite seems to be more prevalent in the oxidized conditions encountered in dry climates. Another hydroxide, lepidocrocite ($\gamma FeOOH$) is less common than goethite, and is most frequently encountered as the orange mottling in gleyed (water saturated) conditions; when dehydrated, it can change to the much more strongly magnetic maghaemite. All these have the very weakly magnetic canted antiferromagnetic type of crystal.

Magnetite and maghaemite, often combined with titanium, are also important soil minerals. Maghaemite is especially important as an indicator of human settlement. Although its chemical formula is the same as that of the low-susceptibility haematite, it has the same crystal structure as magnetite and a similarly high susceptibility.

A number of processes, some linked to man, enhance topsoil magnetism. All involve the development of ferrimagnetic compounds from other forms. Le Borgne (1955, 1960) proposed as a mechanism the alternation of reducing and oxidizing (anaerobic and aerobic) conditions bringing about the conversion of haematite to maghaemite by way of magnetite as an intermediate stage. He postulated a fermentation effect, produced in soil by alternating dry and saturated conditions, and a burning effect in which oxygen was excluded from the ground beneath a fire as it burned, but able to reach it when the fire went out.

Le Borgne considered that susceptibility enhancement by burning was the more significant of these two factors. His evidence in favour of the fermentation effect was less straightforward, but has since been confirmed by other researchers. Earlier attempts to confirm it concentrated on the decomposition of waste material, on the assumption that this had been a major factor in the magnetic enhancement of the contents of ancient rubbish pits, which tend to be particularly strongly magnetic. However, laboratory attempts to reproduce the effect were not successful, and it seems more likely that the pits are magnetic chiefly because their backfilling contains much burnt soil from surrounding occupation areas. Mullins (who, with Tite, has made a major contribution to these studies) proposed a mechanism similar to Le Borgne's, by which the reduction-oxidation cycles which occur under normal pedogenic (soil-forming) conditions lead to the formation of microcrystalline maghaemite or magnetite from weakly magnetic iron oxides and hydroxides. Such cycles would not occur in the deep fillings of pits. There seems no doubt that organic matter is required to support microorganisms which provide the conditions for breaking down the original compounds, but it is not known whether the ferrimagnetic compounds are produced by straightforward chemical crystallization or by way of soil bacteria analogous to those known to secrete magnetite.

An example of the pedogenic process is the substantial enhancement observed in old forest soils; also, measurements on sections in meadowland have shown a clear increase in magnetic susceptibility close to the surface, among and just below the grass roots. A number of experiments and observations indicate that oxidation is important. However, there is no doubt that a very high percentage of any occupied land was originally cleared by the slash-and-burn method, causing thermal enhancement, which has been demonstrated by experiment at the Butser Ancient Farm Research Project.

The strength of the burning effect on occupied sites is such that it seems always to be readily distinguishable when superimposed on any magnetic enhancement present in the generality of soil, and normally on any effect from modern stubble burning, which is very quick and superficial. It has been demonstrated that the burning effect can be enhanced in the presence of organic material, which must provide the necessary initial reducing conditions, and this may in some instances assist the enhancement by fire of the soil in pit fillings.

The burning effect can be readily simulated. One of my lecture demonstrations is to increase the magnetic susceptibility of a sample of soil – usually taken from the nearest flowerbed to the lecture theatre to impress the audience that there is nothing special about the sample – simply by heating it with a butane blowtorch. The strength of the effect, in the absence of any precautions, shows that the conditions for enhancement are by no means critical.

Instruments

First it must be explained why magnetometers cannot be used for all magnetic measurements, in spite of the fact that much of their effectiveness depends upon magnetic susceptibility.

Magnetometers are defined as *passive* instruments, and can only exploit the phenomenon of susceptibility by sensing its effect on the localized shape of the Earth's magnetic field, and its reinforcement or reduction, by distinct archaeological features (Fig. 50). This is demonstrated in Fig. 79, from Aitken (1974), which shows cross-sections of two pits of equal volume, one deep and narrow, the other wide and shallow. The first is equivalent to a substantial bar magnet and is readily detectable, but the second resembles a series of short magnets in which the north and south poles are close together, so that the return flux tends to cancel their effect. This results in a much smaller signal, increasing slightly near the edges where there is an imbalance in the cancelling effect. Taking this illustration a step further, a continuous layer such as the topsoil becomes undetectable.

Magnetic discontinuities in topsoil are better tackled with an *active* instrument, which injects a signal into the ground and measures a response to this signal due to its magnetic susceptibility. The penetration of such instruments is relatively shal-

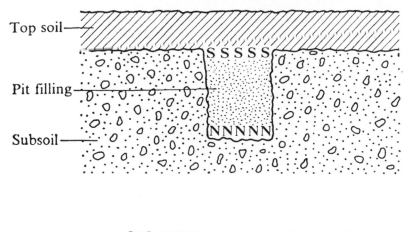

Fig. 79. Two pits of the same volume but different shape. The deep, narrow pit behaves as a bar magnet and gives a much stronger magnetometer anomaly than the shallow pit below.

low because the activating signal is attenuated on its way into the ground, and the return signal on its way out, so that, where r is distance, sensitivity falls off as $1/r^6$, rather than $1/r^3$, as is the case with a magnetometer. This however can be a virtue if one is looking at topsoil, as we shall see. The instruments work on two different principles.

Single loop instruments The standard instrument of this type is that of Bartington, of which the present version is the MS2. The instrument consists of a small electronics box for laboratory or field use, depending on the type of sensor attached. In the field, it is carried in a bag over the operator's shoulder, connected by a lead to the appropriate sensor. Power is supplied to an oscillator circuit at the sensor end which generates a low- intensity (80 A/m) alternating magnetic field in the sensor coil. Material within the influence of this field will change the frequency in proportion to its magnetic susceptibility. The frequency information is returned in pulse form to the MS2, which displays it as a susceptibility value. Readings can be individually taken or continually refreshed. The instrument has an RS232 interface for linking to a

data logger such as the Psion Organiser (compact and reasonably priced), or directly to an IBM-compatible microcomputer. A number of alternative sensors are available for tackling different types of problem.

Laboratory sensors The sensor for precision measurements in the laboratory is the MS2B (Fig. 80). This is a cylindrical coil of 36 mm (1.4 in.) diameter which will take samples in standard tubes of 10 ml or 20 ml volume. With this, accurate measurements of κ or χ can be made with dried and sieved samples. Quick but less accurate measurements can be made on samples straight from the field; χ values will be affected by the weight of contained water, and all readings by any large inclusions. The MS2B has a facility for making dual frequency measurements, which can provide information of archaeological value. Another static sensor is the MS2C, designed for measurements on core samples, and available in a number of diameters.

For dual frequency measurement, the two frequencies used are 0.43 and 4.3 kHz – a separation of a factor of ten. As frequency is increased, susceptibility measurements made with AC become lower than those obtained with a 'DC' field such as that of the Earth, because of magnetic viscosity which slows down the response of magnetic

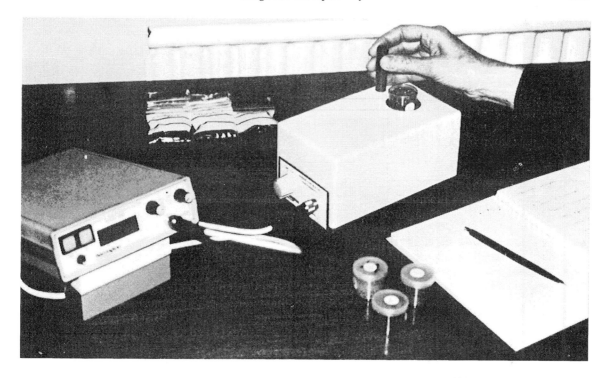

Fig. 80. The Bartington MS2B with dual frequency sensor.

grains to the changing direction of an AC field. All magnetically susceptible natural substances have a characteristic 'susceptibility spectrum', which is a measure of their susceptibility variation with frequency change. The magnetization of materials occurs in discrete 'domains', within which the direction of magnetization is uniform. In relatively large mineral grains, the most stable state is achieved by a multi-domain arrangement, in which adjacent domains have different magnetic directions. Small grains of single domain size (for magnetite, around 50 nanometres or 50 billionths of a metre) are the most stable. There also exist even smaller grains with the property of *superparamagnetism*, which readily align with an applied field, but are unstable when it is removed because of thermal disturbance. Grains around the stable single-domain superparamagnetic boundary display a type of magnetic viscosity resistant to high frequency measurements, so that their apparent susceptibility falls off sharply with increasing frequency.

By sampling the frequency spectrum at two points, dual frequency measurement provides an indication of its form and thus the nature and size distribution of the magnetic minerals in a sample. Material directly weathered from bedrock tends to consist of large, multi-domain grains and shows a low frequency dependence. The processes of reworking by man tend to favour change to the smaller grain sizes, including those with high frequency dependence. Burning has this effect, and it has been observed that fertile, well-drained topsoils, often the product of cultivation, also encourage the precipitation of such ultra-fine secondary iron oxides with relatively high frequency dependent susceptibility.

Thus, frequency dependence can help to distinguish magnetic enhancement due to man from that of natural origin. Magnetic susceptibility measurements over a field may show quite strong variations, but if the frequency dependence is constant and low these are probably due to varying concentrations of natural magnetic minerals in the soil. Sites of human activity will be distinguished as areas of high susceptibility accompanied by increased frequency dependence.

The use of frequency-dependent magnetic susceptibility measurements in archaeology is quite new, but has much potential. It is already increasing our confidence in identifying the settlements of man, sourcing of sediments and correlation of deposits. It is expressed as a percentage by the formula:

$$\%\chi fd = 100 \times [(\chi_{LF} - \chi_{HF})/\chi_{LF}]$$

where χ_{LF} = the low frequency measurement and χ_{HF} = the high frequency measurement.

Field sensors These are the MS2D search loop and the MS2F probe (Fig. 81). The MS2D has a mean diameter of 18.5 cm (7.3 in.) and an effective penetration of about 10 cm (4 in.), while the MS2F, with a tip diameter of 1.5 cm (0.6 in,), is sensitive only to the material with which it is in immediate contact (or immediately surrounding it – see below). Both sensors are used with a standard handle assembly of adjustable length, to which the oscillator electronics are attached. As there is no control over the condition or mass of the ground sampled in field measurements, these

sensors are calibrated to give volume susceptibility (κ) measurements.

The MS2D is the standard sensor for fieldwork. It requires intimate contact with the ground to give full readings, because there is a rapid fall-off with distance. Readings are roughly halved at 2 cm (0.8 in.) separation and reduced to a quarter at 4 cm (1.6 in.), and 10 cm (4 in.) is the response limit. The MS2D is thus fully effective for clear, smooth ground and for ground with relatively thin grass or a young cereal crop, but care has to be taken with furrowed ground, and tussocky vegetation can defeat it. One of the design objectives of the MS2F was that it should be small enough to be pushed through such vegetation, but even this detector can be defeated by the root mass. A solution is to drop it into shallow holes made with a 2.54 cm (1 in.) diameter corer: its sensitivity is not confined to its tip, but extends along the narrow diameter so that readings of the hole sides are obtained which are closely comparable with those obtainable with the tip (Fig. 82). The MS2F was, in fact, designed primarily for the

Fig. 81. (Left) The Bartington MS2D search loop. (Right) The MS2F probe.

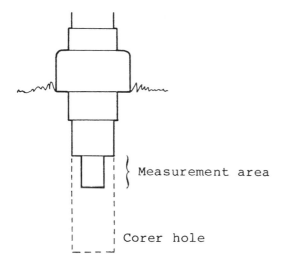

Fig. 82. Using the MS2F probe in a 2.54 cm (1 in.) cored hole. The hole does not have to be deep, but its bottom and any disturbed soil within it should be well clear of the probe so that only the side of the hole is sensed.

detailed examination of flat surfaces and sections, an application which will be discussed later.

The prototype for a true shallow borehole sensor (the MS2H) has been tested. The diameter of this required quite a large hole, and a version suitable for a narrow and more easily drilled hole is being designed.

Transmitter-receiver instruments The use of this type of instrument for resistivity, or rather conductivity, measurement has been discussed in Chapter 2. A remarkable facility of such instruments is that they can be made to respond to either the magnetic susceptibility or the conductivity of the soil. We have seen that the response which is out-of-phase with the transmitted signal is used for conductivity measurement. In the case of magnetic susceptibility, the instrument responds to the varying magnetization caused by the primary field in any susceptible compounds within the ground. This response is in-phase, with only a small out-of-phase component due to magnetic viscosity. The instrument discussed in Chapter 2 as being effective for conductivity measurements was the Geonics EM38, which can also be used for magnetic susceptibility. Such instruments which normally operate with a continuous sine wave were named Slingram when

developed in Sweden in the 1930s (Scollar *et al*, 1990).

Much attention has been given to the optimization of this type of instrument by Tabbagh; his currently most developed version, the SH3, has a coil orientation of 35 degrees from vertical, 1.5 m (5 ft) separation, and an operating frequency of 8 kHz (Tabbagh, 1986). Carried at a height of 15 cm (6 in.), this has a susceptibility depth penetration of 70 cm (28 in.), which could be increased by using a larger coil spacing though to some detriment of spatial resolution and convenience of operation. Thus instruments of this type have much deeper penetration than the Bartington, although in practice this can actually be a disadvantage, as will be discussed below. Instruments in this class are suitable for continuous recording, with the reservation that height control is more critical than with magnetometers.

The first instrument based on this principle to

Fig. 83. Mark Howell, the inventor, using the SCM or 'Banjo'. The transmitter coil and power source are in the vertical box at the back, and the circular horizontal receiver coil is at the front.

be used in archaeology was the SCM (Soil Conductivity Meter) of Howell, already mentioned in Chapter 1, which actually proved to respond to soil susceptibility (Fig. 83). This instrument, nicknamed the 'Banjo', operated at 4 kHz and had transmitter and receiver coils in a perpendicular configuration so that there was no interaction between them unless the transmitted signal was distorted by variations of ground susceptibility. In fact, there are a number of pipe finders which are very similar. The original Fisher M-Scope, for instance, has been found very effective for detecting soil susceptibility changes.

Pulsed induction meters All magnetic susceptibility instruments are highly sensitive to metal, and therefore double as metal detectors – which is sometimes a disadvantage. The ingenious pulsed induction principle was actually developed by Colani as a metal detector, but was also found to be sensitive to the magnetic viscosity of iron compounds in soil, a property discussed above in connection with dual frequency susceptibility measurement. Viscosity has been assumed to be approximately proportional to the susceptibility, and thus an acceptable substitute when used for comparative measurements in the field. However, as we now know that short-term viscosity effects tend to be low in grains derived from bedrock, but high in those modified by man, it seems that pulsed induction is preferentially sensitive to soils affected by human occupation.

The basic principle of pulsed induction makes use of the slow decay of induced magnetic fields. In its original form, the instrument antenna consists of two concentric coils, a transmitter and receiver (Fig. 11). A strong DC pulse, lasting about half a millisecond, is sent through the transmitter. The receiver is switched off during this pulse and is therefore unaffected by it, but is switched on about 50 microseconds later in order to detect any signals transmitted back from the ground. These can be due to electric currents (eddy currents) generated by the transmitter field in buried metal, or viscous magnetic decay in soil. The great advantage of this principle over continuous signal metal detectors is that the risk of interference between the transmitter and the receiver is avoided, and thus the activating signal can be very strong and the receiver very sensitive. There is the further benefit that the effectiveness of the instrument does not depend on rigidity of construction, so the coils can readily be made quite large to aid penetration.

Modern switching circuitry makes it possible for a single coil to be used for transmission and reception in most pulsed induction meters. Although it was once hoped that this principle could equal or exceed the effectiveness of magnetometers for detecting archaeological features, the fall-off in sensitivity with depth, common to active instruments, proved to be an impenetrable barrier and most pulsed induction meters are now designed specifically as high sensitivity metal detectors. They are tuned for maximum efficiency for this purpose, and are not highly sensitive to soil variations. In view of their possible advantages, their revival for this type of work would seem to be desirable.

*

Much of the background to the discussions of soil properties in the above sections has been derived from Mullins (1977), Tite and Mullins (1971), Tite (1972b), Tite and Linington (1975) and Longworth and Tite (1977). More reference will be made to some of these in the Supplement. It is possible to make a number of other magnetic laboratory measurements to obtain further information on the nature and origin of magnetic minerals contained in soils. These cannot be pursued here, but are comprehensively described by Thompson and Oldfield (1986).

Field survey It has long been realized that enhanced phosphate levels can occur in soils once occupied by man; this will be discussed in Chapter 5. The phosphorus is derived from waste products such as excreta, and from bones, and is remarkably tenacious. Following the studies of Le Borgne, it was soon appreciated, notably by Tite, that areas of former human occupation could also be broadly defined by means of magnetic susceptibility measurements on surface soil. It has since become clear that this evidence is not always greatly diffused by centuries of ploughing, and that site structure and usage can be preserved in astonishing detail in the topsoil, or the 'plough-zone' as it is sometimes called. Prehistoric man, like us, lived his life on the surface of the ground and did not expend the effort to dig into the bedrock unless he had to; yet archaeologists have concentrated upon such subsurface features because the topsoil has been assumed to be too

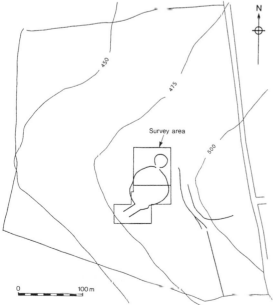

Fig. 84. (Above) Parched crop mark of the Tadworth site, 16 July 1975. Epsom Racecourse is in the background. (Left) Interpretation, with survey grids superimposed. The survey considered here is the upper 60 m (195 ft) square. (*Crown copyright.*)

thoroughly mixed and transported by ploughing, worms, weathering and gravity to preserve any coherent information. Often this is the case, but often not, and on some sites stripping the topsoil is stripping away archaeological information. The main criterion for the more detailed preservation is that the site should be fairly level; even if it is not, much information can still be retained in the soil (Gingell and Schadla-Hall, 1980).

The Tadworth experiment In 1977, N. S. Littler and myself, in collaboration with the Royal Commission on Historical Monuments Air Photographs Unit and Surrey University, made a study

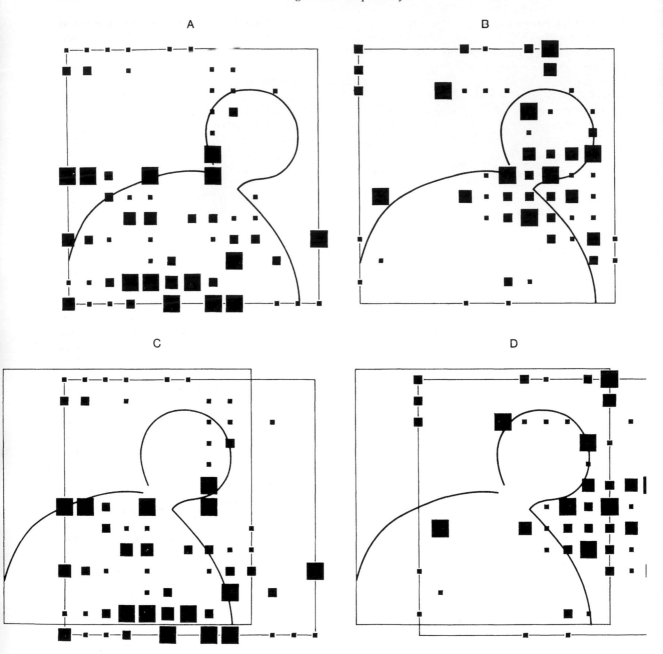

Fig. 85. Tadworth. (A) Pulsed induction survey. (B) Phosphate survey. Ground reading interval = 5 m (15 ft). Both surveys are manually plotted using four increments of symbol size at intervals of approximately half the standard deviation above the mean + ⅙ s.d. A: Mean = 20, s.d. = 6 arbitrary instrument units. B: Mean = 188, s.d. = 62 ppm. (C and D) The data plots shifted uphill to compensate for downhill soil drift.

of a site which appeared to be an Iron Age 'banjo' enclosure on Upper Chalk at Tadworth in Surrey (see Groundwell Farm, page 139, for a definition of the banjo form – not to be confused with the Banjo instrument!). There were two objectives: to use geoprospecting methods to amplify the evidence of air photography, by which the site was discovered (Fig. 84), and to ascertain whether phosphate survey could be replaced by magnetic susceptibility, which is much faster, on the assumption that it would have been enhanced by the fermentation effect associated with the waste materials that are responsible for phosphate enhancement. Resistivity was able to show the ditches as weak positive anomalies while a magnetometer survey was unresponsive; from this it was concluded that there had been heavy plough erosion and that the ditches and any pits, which may have been slight from the beginning, had probably been truncated down to primary silt level. For the electromagnetic survey, a pulsed induction meter was actually used because suitable accurate field susceptibility instruments were not then available. The two surveys were within a 60 m (200 ft) square covering the northern part of the enclosure, which included a curious annexe that looked suitable for a house site.

The results, plotted in Fig. 85A and B, were quite unexpected. Rather than defining the same areas, the two surveys showed almost zero correlation. The electromagnetic survey was more responsive to the main enclosure, whereas a distinct grouping of high phosphate readings was more closely associated with the annexe. It was therefore concluded that the susceptibility enhancement was probably due mainly to the residues of fires in the main enclosure, and that the annexe was probably an animal compound. Thus the hoped-for conclusion that phosphate survey could be replaced by susceptibility had to be abandoned, but in favour of the much more exciting discovery that the two techniques could define areas of differing activities and produce enhanced interpretative detail without excavation, even on sites where individual features had been heavily eroded.

The lack of exact coincidence between the areas of enhancement and the site boundaries was ascribed to soil drift, which one would expect on a ploughed site of great antiquity, and the site was published (Hampton *et al*, 1977) with these conclusions. Two aspects of this interpretation,

however, caused some unease. As the contours in Fig. 84B show, the average slope of the northern part of the site is about 1:20 in a direction a little north of west, so that the assumed southern drift of the phosphate would actually be slightly uphill, which seemed highly unlikely. It was also noticed that the shape of the western side of the enhanced area was similar to the re-entrant angle between the main enclosure and annexe ditches. If the phosphate survey is adjusted to the best fit into this angle, the direction of movement necessary is uphill exactly along the line of maximum slope; moving the susceptibility survey similarly gives a better coincidence with the main enclosure, with some scatter in the general area of the phosphate concentration (Fig. 85C and D). A number of strong responses coincide with the ditches.

Thus, by assuming a downhill soil drift of 15 m (50 ft), the interpretation is radically changed. The phosphate enhancement now seems to be due to the judicious deposition of waste outside the enclosures; the annexe becomes strikingly clear of both phosphate and magnetic enhancement, and could now perhaps be interpreted as a clean area for storage – or even a well-maintained house area, as originally guessed. The 15 m (50 ft) drift seems alarmingly large, but data obtained by Taylor from one of his experiments on a similar lithology seem to indicate that the rate of drift was about 0.7 m (2.3 ft) in 35 years of ploughing on a 2° slope (Taylor, 1979). Adapting this to the slightly steeper slope at Tadworth would give a movement of about 1 m (3.3 ft) in 35 years, or 2.86 cm (1.1 in.) per year, which could mean that the field had been ploughed about 525 times. It would be interesting to compare this with local records. One of the several significant points to emerge from this exercise is that, throughout all this movement, the phosphate area particularly seems to have retained its coherence and outline sufficiently well for its fit to the ditch angle to be recognizable: continental drift (which once seemed equally incredible) in miniature!

At Butser Ancient Farm Project, Peter Reynolds has established a long-term experiment to test soil drift and artefact movement. Magnetic simulated pottery sherds were implanted in known patterns; the area has been repeatedly ploughed and the samples relocated by magnetometer each time. The patterns are retaining their coherence well but an overall downhill drift is beginning to be noticeable.

General surveys Non-detailed surveys can be used for the prediction of areas worthy of closer attention, either by susceptibility or magneto-meter measurement, or by any other available technique. Conversely, they can be used after a magnetometer survey, to confirm that the limits of a site have indeed been reached.

Measurements on a grid as coarse as 30 m (100 ft) mesh, or even greater, can be effective for defining the broad limits of large occupation areas. In Fig. 105, seven measurements over an area of about 5.5 ha (13.5 acres) were sufficient to establish clearly the focus of a settlement. Another minimal survey at Braughing, Herts, was most effective in broadly defining the south-ward extent of the Roman town there. In less than a day, 7 ha (17.3 acres) were examined: 168 susceptibility spot readings showed that high values were concentrated in the northern part of the area, and these were supported by seven lab-oratory readings on collected samples and a flux-gate gradiometer scan which confirmed the existence of archaeological features within the high- susceptibility area. The range of laboratory susceptibility readings was 8.5 to 58 x 10^{-8}SI/kg.

Detail recorded in topsoil At first, it seems incredible that individual features can survive in top-soil, which in many cases has been long under the plough, but the evidence is inescapable.

Bowen (1975) was the first to draw attention to linear soil marks, on air photographs taken over the chalk country of Wessex. These produced no corresponding marks under crop, which suggested that there was no penetration into natural and that the marks were caused by bands of top-soil darkened by relatively high water retention. Tests confirmed that at least some of the marks were not underlain by features.

There is a famous field beside Micheldever Wood, north of Winchester, which has probably been photographed from the air more than any other in Hampshire. Adjacent to it in the wood, and now obliterated by the M3 Motorway, is the Iron Age banjo enclosure discovered by mag-netic scanning; and in the field itself, opposite the entrance to the enclosure, photographs taken when the soil was bare revealed parallel dark lines that may be associated with it. Excavations by Peter Fasham, in the course of the M3 archae-ological programme, showed that the marks were underlain by the last remnants of truncated ditches, or in some places nothing at all where the ditches had been eroded away completely. The field had been ploughed since the late eighteenth century and was estimated to be losing its chalk bedrock at a rate of 5 mm (0.2 in.) per year; however, it is almost level so that the soil marks, which must represent remnants of the ditch fill-ings, still remain *in situ*.

Taylor (1979) analysed the soil of these marks, in comparison with the soil away from them. The marks showed no increase in humus content, but there was an increase in clay and the ratio of small to fine particles. These would tend to increase the water retentivity of the soil, and could bring about the dark marks in the right conditions of water balance. Taylor did not have access to mag-netic susceptibility measurements, but these would almost certainly have been sensitive to the enhanced content of clay. He felt that, although the dispersal was remarkably small, continued ploughing where the ditches had been finally removed would lead to increasing dilution with chalk and quite rapid dispersal, so that the marks would disappear. Similar results were obtained over a ditch nearby at Chilbolton. Another set of similar marks a few miles away at Portway, Andover, was tested by excavation which showed no underlying features. Resistivity tests by myself showed a very slight fall in value, close to the limit of detection, over the marks, which was in keep-ing with Taylor's findings. Such features should be detectable by susceptibility. At Sutton Hoo, a number of ditches of the prehistoric settlement which were visible from the air but undetectable by other instruments seem to be detectable by susceptibility, but this work has not yet been fully evaluated.

The first magnetic susceptibility survey pur-posely designed to search in detail for possible features preserved in the topsoil was on the site of the small henge monument of Coneybury, near Stonehenge. Because of its dramatic success, it has received much exposure in publication but cannot be ignored here. It provides a good start-ing point for discussing the mechanisms that make possible both the preservation and the de-tection of archaeological features in the topsoil.

Coneybury was first discovered as a crop-mark from the air, and partly excavated in 1980 by Julian Richards for the Trust for Wessex Archae-ology, with the technical support of the Ancient Monuments Laboratory. It proved to be a simple monument of the henge type, with a substantial ditch and single entrance, no stones or major post holes, but a group of shallow pits around its centre. The ditch was first located with a conventional fluxgate gradiometer survey which was followed by a fairly intensive (0.7 m (2.3 ft) diagonal squares) magnetic susceptibility survey with a prototype of the MS2D loop sensor (Fig. 86, left). One of the attractions of the site for this first experiment was that it was almost completely level, with just a hint of hollowness, so that pro-

blems from soil drift were not expected to be serious. The average reading for the surrounding field was ascertained, and the differences of readings from this were plotted as black squares above and open triangles below, varying in size according to the difference. This seemed the clearest way of presenting the rather 'noisy' data.

Most immediately obvious was a grouping of high readings on the main axis of the henge, which was interpreted as the possible site of a central fire. A sprinkling of high readings ran along the ditch, and there was a fair concentration at the entrance. Between the putative fire and the entrance, low readings predominated, forming diagonal bands along the line of modern ploughing. The soil was relatively shallow here, and the effect was clearly due to the plough cutting into the natural chalk, which had diluted the topsoil and reduced its susceptibility.

That the central grouping of high readings did indeed represent a fire was confirmed by careful hand excavation of the topsoil, in 1 m (3.3 ft)

Fig. 86. Coneybury henge. (Left) Computer plot of the magnetic susceptibility survey with the MS2D loop sensor (framed) superimposed on the interpretation of the fluxgate gradiometer survey of the ditch (solid outline). The side lengths of the solid squares and open triangles increase in steps of 1.6×10^{-5} SI above and below the site background level. The squares are in five steps between 34 and 42 units; the triangles in four steps between 29.2 and 22.8 units. (Right) Plot of the weight of burnt flint in the topsoil per square metre of the excavated area, over the range 0-492 g/m².

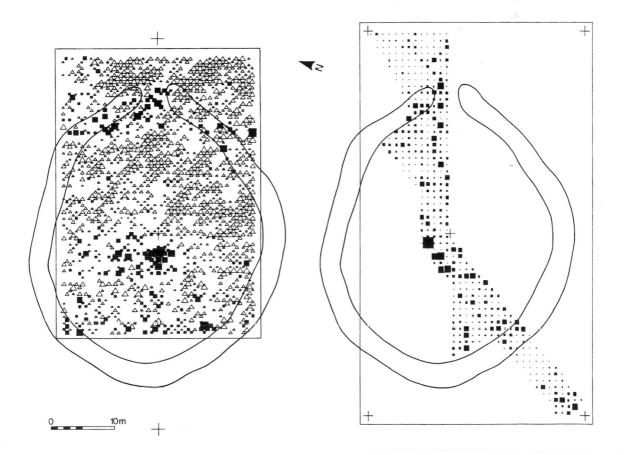

0   10m

squares, which showed a concentration of burnt flint in this area (Fig. 86, right). Nothing was found on or in the natural chalk beneath, showing that the evidence was contained entirely within the soil, yet had retained its integrity for four thousand years. Some phosphate enhancement was also found in this central area, so this fire may well have been used for rituals of feasting or cremation – the parallel with the central feature of the much grander henge at Stenness (Fig. 72) is inescapable. There are two possible explanations for this survival. One is that all the enhanced soil had remained in the tilth, and that there had been no net soil movement due to ploughing. In turning a furrow (Fig. 87), the main effect of the plough is to move the soil to one side, but this can be annulled by ploughing in opposite directions. The experiments at the Butser Ancient Farm Project are certainly showing that overall soil movement can be very small after repeated ploughing, and we have already seen at Tadworth that, even when there is substantial overall movement due to slope, the shape of the enhanced deposit appears to have been retained.

However, over the western half of the henge the natural chalk is relatively deep and some relict ancient soil, which would have been affected by the fire, underlay the modern ploughsoil. Recent deep ploughing, as evidenced by the dilution lines already mentioned, could have brought up some of this soil, reinforcing the magnetic susceptibility at the focal point, even if there was an

Fig. 87. The action of deep ploughing in bringing small amounts of an archaeological feature to the surface.

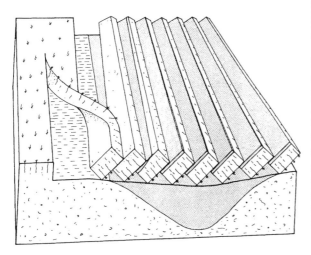

overall tendency for dispersion. Thus the clarity of the Coneybury fire may be due to recent deep ploughing which, if continued, might have obliterated the evidence, as Taylor suggested could happen at Micheldever Wood.

An appearance of quite dramatic dispersion is seen on some air photographs, which show linear soil marks turned into zigzags by the plough; but Wilson (1982) points out that this effect is superficial, and produced by the spreading of very small amounts of material (Fig. 87). This is shown by the fact that, rather than increasing, the dispersal is effectively annulled each year by incorporation of the spread material into the soil and replaced by a fresh mark if the plough continues to cut deeply enough. In magnetic terms, this process is likely to be represented by relatively strong enhancement closely associated with the location of the feature, surrounded by a region of much slighter enhancement. Obviously, however little material is shaved off each year, the buried feature in such conditions is a finite resource. Fortunately, the vogue for deep ploughing is now waning in Britain, as indeed is the overall expansion of cultivation, so many such features may be saved. Shallow ploughing is increasingly incorporated as a stipulation in scheduled monument management agreements by English Heritage.

The success of the work at Coneybury led to the inclusion of magnetic susceptibility survey in the broader Stonehenge Environs Project, in which it proved valuable for indicating focal areas of prehistoric occupation sites initially located by fieldwalking, as well as revealing some evidence of site structure, again with possible fires, inaccessible to the magnetometer (Entwistle and Richards, 1987).

Sometimes the findings of magnetic susceptibility survey are not readily explicable, at least in our present state of knowledge. A group of small prehistoric fields was investigated in 1986 at Dainton, Devon, by the English Heritage Central Excavation Unit, under the direction of George Smith. These had a shallow soil on limestone, and were still separated by low banks, shown in Fig. 88. Magnetic susceptibility and phosphate samples were taken from the topsoil on a 5 m (16.4 ft) grid, and shading density plots by Nick Balaam revealed the distinctive distributions shown.

Phosphate levels were high, especially in the

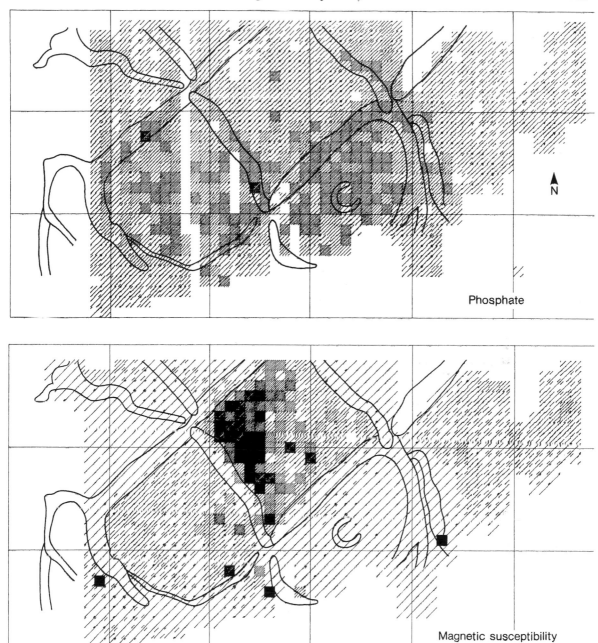

Phosphate

Magnetic susceptibility

Fig. 88. Dainton, prehistoric fields. Phosphate is in
500 ppm steps up to 2,000 ppm; magnetic susceptibility is in steps of 10×10^{-8} SI/kg up to 40.

two southern fields, the eastern of which contains the foundation of a circular house. Susceptibility was greatest in the central northern field and for that reason excavation was concentrated in it, although nothing distinctive was found. Stone imported to the site was studied by Brian Selwood, who discovered that the field system was littered with small pieces of burnt igneous dolerite, of which the nearest known outcrop is 2.5 km (1.6 miles) away. The dolerite was used as temper in Late Bronze Age pottery on the site, but was considered much too widespread to be connected with the pottery alone. It was therefore suggested that the dolerite, rich in ferromagnesium minerals, might have been used as a fertilizer – a remarkable sophistication for Bronze Age farmers if true.

The burnt dolerite probably explains the susceptibility enhancement of some of the fields. The exceptional values in the central field might have been caused by its being used for the preparation of the dolerite, or possibly bonfire pottery clamps, at some stage in its existence. Stirring of the soil by subsequent ploughing could have dispersed any direct evidence of such usage.

An interesting detail is the tendency of phosphate to concentrate along the field banks. One can only surmise about the meaning of this. Was manure piled along the edges of the fields?

The excavators remarked on the surprising lack of susceptibility enhancement around the house. Phosphate was high there, and it has been suggested that high phosphate may suppress magnetic susceptibility. However, the evidence from the south-west field, where there was enhancement of both phenomena, did not support this. Susceptibility suppression is at present only certainly known to be caused by the anaerobic conditions of gleyed (water-saturated) soils. These conditions favour the reduction of iron compounds from the ferric to the relatively weakly magnetic ferrous state, apparently with the help of bacteria. It is important to consider the possibility of gleying when interpreting a magnetic susceptibility pattern.

Magnetic susceptibility features in grassland We have assumed much involvement of plough action in generating or spreading the evidence of past activities in the topsoil. Good responses can be obtained on grassland which may never have been ploughed, and there seems no doubt that here the plough is largely replaced by earthworms. Charles Darwin (1883) summarized their influence thus in his *Vegetable Mould and Earthworms*:

> When we behold a wide, turf-covered expanse, we should remember that its smoothness, on which so much of its beauty depends, is mainly due to all the inequalities having been slowly levelled by worms. It is a marvellous reflection that the whole of the superficial mould over any such expanse has passed, and will again pass, every few years through the bodies of worms. The plough is one of the most ancient and most valuable of man's inventions; but long before he existed the land was in fact regularly ploughed, and still continues to be thus ploughed by earth-worms. It may be doubted whether there are many other animals which have played so important a part in the history of the world, as have these lowly organised creatures.

Thus, at least on *in situ* soils supporting worms, there should be no danger of more recently developed soil masking the occupation effects of ancient man, which should be well dispersed vertically – and, presumably, also to some extent horizontally, although we lack control data on lateral dispersion on turf-covered sites. The upward movement, probably by worms, of material from a hearth has been confirmed by Yates (1988), using a Bartington MS2F probe for detailed examination of sections; and, quite remarkably, a prehistoric site at Peterborough was detected by phosphate enhancement of topsoil separated from the archaeological layers by 0.5 m (20 in.) of alluvium (Craddock *et al*, 1985). There seems no reason why magnetic susceptibility should not be equally responsive in such conditions if the movement is due to worms; but the phosphate transport may be assisted preferentially by plant roots. Research is urgently needed to discover the mechanism and limits of such vertical diffusion, as these techniques could provide at least a partial solution to the widespread problem of locating alluvium-blanketed sites. More measurements are needed where suitable excavated sections are available. Whatever is achieved in this direction, there will remain sites where gross soil movement has occurred, or peat has developed, and a borehole strategy is needed. Without a borehole susceptibility probe, such measurements can be made with separate extracted samples, using the MS2B laboratory sensor, with the diagnostic power of the dual frequency facility. Alternatively, the MS2F probe can be dropped into holes at fixed increments of depth.

Magnetostratigraphy: landscape on the move

This is the magnetic study of cores and sections of accumulated deposits. Although not a direct method of site prospecting, this subject falls legitimately within the remit of this book because it can provide evidence of human occupation by way of the soil eroded from a landscape.

We have seen above that the topsoil varies greatly in mobility. On a level site, it may be expected to remain broadly *in situ* indefinitely. On a gentle slope, plough action loosens the soil and allows it to be slowly moved downhill by the action of rain, frost and gravity. On steeper slopes, major erosion can take place, sometimes to the accompaniment of gullying; the inwash, *en masse*, of topsoil from a steep field into a dry valley has actually been witnessed by Michael Allen while he was excavating in the South Downs. In situations of this kind, or other sites with deep stratigraphy, occupation may have been established at a succession of levels, often in a deposit showing little colour differentiation; susceptibility measurements can be useful during excavation for locating the occupation levels and features within them, especially those associated with burning.

Such effects have been put to good use in archaeology for establishing the history of landscapes adjacent to both dry and lake-filled valleys. Most striking is usually the effect of the initial clearance of the land by slash-and-burn, which both enhances the magnetic susceptibility of the soil and makes it vulnerable to erosion. Subsequent periods of alternating stabilization and further clearance can be recognized, and illuminated and correlated by pollen and other environmental studies. A team from Liverpool University is at present involved in such a coring programme in Cumbria, to discover lake basins settled by prehistoric man. Positive results from Little Hawes Water were followed by a wide ranging surface magnetic susceptibility survey of the basin which identified three compact areas

of enhancement likely to represent occupation sites. These in their turn have been cored to show that the enhancement is present in depth, rather than superficial as it would be if due to modern activity. The next stage was a test excavation of one of them, which is producing evidence of occupation.

Yates (1989) has found magnetic susceptibility a good measure of the intensity of usage of occupation levels on settlements ranging from the Koster site, Greene County, Illinois (a classic reference site for American prehistoric chronology where a long sequence of episodes of occupation is clearly defined by intercalated colluvium) to modest and relatively short-lived English sites. At Koster, layers representing transient use and those representing prolonged and settled occupation were difficult to distinguish on the basis of appearance or artefact density; but the latter gave higher susceptibility measurements because of accumulated magnetic enhancement. At an English site, low susceptibility values for a layer containing charcoal contrasted with high values for a hearth and its surroundings on another part

Fig. 89. Drayton Cursus. Magnetostratigraphy of the ditch silting. The dates are derived from archaeomagnetic measurements on the silt.

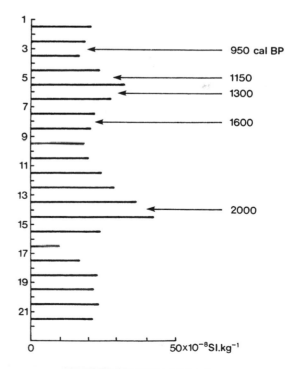

of the site. This showed that the charcoal had been scattered from elsewhere rather than burnt *in situ*, and therefore did not represent one of the main occupied parts of the site.

An illuminating study has been made of one of the ditches of a Neolithic cursus at Drayton, Oxfordshire, on the gravels of the Thames valley, where later flooding had filled the ditch with alluvium (Fig. 89). It was possible to date the accumulation of this by archaeomagnetic measurements. Parallel susceptibility measurements showed peaks in the Late Iron Age and early medieval period, indicating major episodes of run-off upstream, probably due to clearance and increased cultivation. The Thames is well placed to show such effects clearly, because its headwaters lie in the Cotswold region, where soils are particularly magnetic. The use of dual frequency measurements, not available at the time, would probably have given additional information on the extent to which the soil had been cultivated before it was eroded.

The examples above are straying from prospecting to site analysis, an area in which magnetic measurements are making increasing contributions in archaeology.

Surveys over igneous bedrock We have already seen that the strong original thermoremanent magnetization of the rock in areas of igneous geology can make ordinary magnetometer surveying difficult or impossible. In such situations, magnetic susceptibility may be more successful because this type of measurement does not respond to the existing magnetic fields of the rocks and, because of its shallow penetration, is more sensitive to even thin soil cover than to the susceptibility of underlying rocks.

A good example is the survey of the Amerindian site at Pointe de Caille, at the southern end of the Caribbean island of St Lucia (Fig. 90). The site is on a small, grassy peninsula, and had been recognized by the erosion of midden deposits along the shore; its extent needed to be discovered as a preliminary to rescue excavation. The pace of the erosion was demonstrated by its encroachment on the line of a road made during the Second World War to service a radar station at the end of the peninsula.

The island is highly volcanic, and the peninsula is formed of igneous pebbles and boulders in a soil matrix. A test survey with the fluxgate gradiometer produced responses showing some rough coincidence with the occupied area (bottom right plan), but was dominated by the jumbled response to the magnetic stone. Some 'quieter' areas indicated either gaps in the stone spread or increased soil depth. A thorough resistivity survey might have been effective, but there was no time for this.

Magnetic susceptibility, measured on a 1 m (3.3 ft) grid with a Bartington MS2D loop sensor (top left plan), defined a more coherent main area which had already been seen to contain a hearth and a mortar visible on the surface; its north-west boundary coincided with a still visible stony bank which may once have encircled the site. This picture was clarified by filtering (top right plan) which showed the highest readings around the periphery of the site, perhaps due to accumulation of rubbish behind this bank; a roughly concentric inner pattern may also be significant. An interesting technical point was the substantially greater enhancement of the susceptibility along the edge of the cliff where the erosion is active. Exposure to the warm wind that constantly blows against this shore and the alternating rain and dryness of the Windward Islands would seem to provide very good conditions for accelerating the processes of pedogenic enhancement.

Subsequent excavations by a team from Vienna State University, directed by Herwig Friesinger, confirmed that the site had been correctly defined by the survey, and finds and plans are now housed in a small museum specially built nearby. Pointe de Caille, and other sites with igneous interference, would probably benefit from dual frequency measurement, especially if adapted to a field probe. The archaeological picture should be sharpened by the ability of this technique, and perhaps also pulsed induction, to distinguish the small particles associated with human activity from multidomain grains directly derived from the igneous petrology.

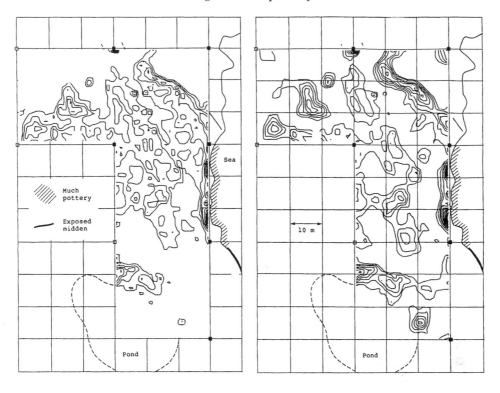

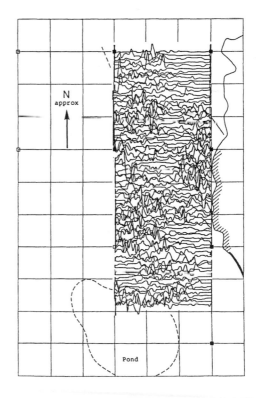

Fig. 90. St Lucia, Windward Islands. Magnetic susceptibility compared with magnetometer survey over igneous erratics. The site was accurately defined by the susceptibility survey. (Top left) Raw data from the mean to the maximum, $130–344 \times 10^{-5}$ SI, contour interval half the standard deviation = 23 units. (Top right) Readings treated with a band-pass filter between limits of 1.5 and 4.5–6.0 m (5 and 15.5–20 ft) ground radius. Values plotted are from the mean to the maximum of the residuals (6–233 SI units), with a contour interval of 23 units. (Right) Fluxgate gradiometer survey, sensitivity 25 nT/line interval. Note the random directions of the peaks caused by the boulders.

Chapter 5
Other Methods

The techniques discussed in the previous chapters are those that are routinely used. Not the least significant of their advantages is speed. This chapter briefly reviews some emerging techniques, and others whose usefulness may always be relatively limited. They are placed in order of importance as it appears at the present time.

Ground penetrating radar

This technique has the very attractive ability to provide subsurface profiles. Although first applied in archaeology in the USA in the early 1970s, its use has nevertheless been limited, especially in Britain. In practice, the records produced have been quite difficult to interpret and its effectiveness has been limited in wet soils. Recent improvements are likely to widen its usefulness although the equipment is expensive and complicated, and is therefore best hired, which again is expensive because an experienced operating team is required.

Pulses of electromagnetic energy are sent into the ground from a transmitter mounted under a small trolley. Reflections are obtained from interfaces between layers and objects of contrasting electrical and magnetic properties, whose

Fig. 91. Ground penetrating radar in use by Jon Glover and FEP Ltd at Woodhenge.

depth can be estimated from the time taken for the reflected wave to return. The technique is closely analogous to seismic reflection but, because of the high frequencies (short wavelengths) used, it is more suitable for shallow probing. A useful frequency is 250 MHz (250 million oscillations per second), which gives a good compromise between penetration and resolution. The trolley is moved steadily over the site and a continuous record is obtained. Fig. 91 shows a small radar detector being operated at Woodhenge, and the record emerging from the plotter.

Ground radar signals tend to be heavily attenuated by moisture, which is one of the reasons why the method has had limited use in the temperate climate and moist soils of Britain. In con-

Fig. 92. Radar profile of an ancient Japanese dwelling buried under a pumice layer. The delay time for the return signals is calibrated in nanoseconds (billionths of a second).

trast, it has been most effective in Japan, where many archaeological sites are covered by a blanket of uniform and well-drained pumice spread over the country by volcanic eruptions. The clarity of response in such conditions is exemplified by Fig. 92, a survey made with a YL-R2 system produced by the OYO Corporation (Imai *et al*, 1987). Although excellently clear, this profile does show one of the problems of ground radar: multiple reflections which, together with lateral reflections, can produce confusing effects with more complicated sites.

Improvements have recently been achieved by the application of increased computer sophistication in radar imaging. An example is the work of Geospace Consultancy Services Ltd, who have used such computer processing combined with the extremely effective OYO equipment to produce uncomplicated cross-sectional images of deep stratigraphy at York. These have been very helpful for site evaluation and excavation plan-

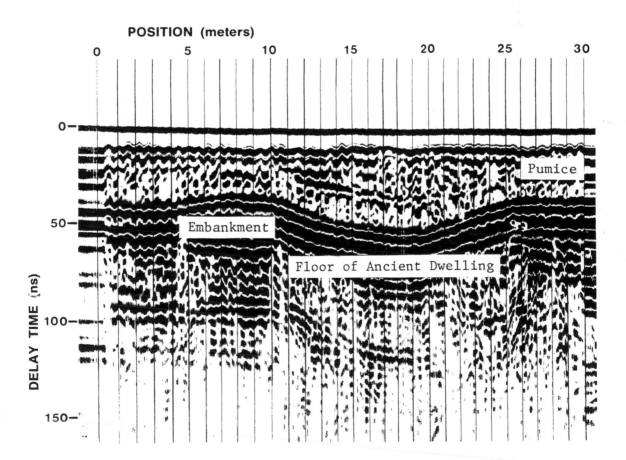

ning in cramped and difficult conditions (Stove and Addyman, 1989). One of their images in this *Antiquity* publication shows a section scaled to a depth of 8 m (26 ft). This is interpreted as showing a Roman wall foundation at 6.4 m (21 ft) with visible offsets, and a number of stone surfaces overlain by nearly 3 m (10 ft) of Viking-age organic layers. These are cut through by a medieval well, in which a barrel lining can be discerned. Such images depend upon the use of colour for their full clarity.

Work of comparable quality has been produced by Oceanfix International Ltd at Sutton Hoo, where Michael Gorman previously carried out pioneering work in high-resolution radar imaging (Gorman, 1985; Carver, 1986). The use of ground-based radar in archaeology is discussed at some length by Weymouth (1986).

An article also in the issue of *Antiquity* containing the report on York describes extremely valuable data obtained by radar mounted on a Space Shuttle. This revealed a complex of former river channels, heavily alluviated and finally obliterated by the wind-blown desert sand on the border between Egypt and Sudan: the desert appeared to be peeled away. Ground investigation has located Acheulian tools within some of these valleys, dated to about 141,000 BP by the uranium-thorium method (McHugh *et al*, 1989).

Allied to this type of imaging is the use by the United States Minerals Management Service of high-resolution seismic reflection profiling to evaluate the archaeological potential of the seabed on the continental shelf of the Gulf of Mexico, which was dry land in prehistoric times (Stright, 1986). The method does not have the resolution to detect archaeological sites directly, but clearly pictures such geomorphic features as fluvial systems, bays and lakes which would have been preferred areas for human occupation, locatable by coring.

Geochemical methods

By far the most exploited of these methods is phosphate detection, which has been mentioned a number of times in this book. Phosphates, originally deriving mainly from natural apatite present in nearly all rocks, are taken up by living things which concentrate them in their bodies and excreta. It has been estimated (Proudfoot, 1976) that the activities of a population of 100 people

would deposit some 124 kg of phosphorus annually. Re-deposited in organic form, the phosphates tend to bind strongly and in effect permanently to the clay-size fraction of the soil, changing slowly to even more stable inorganic forms. Modern phosphate treatments are designed to disperse rapidly and do not normally seem to affect the detectability of ancient phosphate, even if they enhance the overall background level.

The most useful modern practical summary is that of Craddock *et al* (1985), while one by Proudfoot (1976) concentrates more on the underlying chemistry. The first paper includes a description of the sampling and measurement techniques developed by the British Museum Research Laboratory, which have been improved and refined over the years. The method is only claimed to be semi-quantitative, but its relative simplicity suits it well to operation on site. Samples consisting of one gram of soil are dissolved in hydrochloric acid, and a small amount of the solution added to a molybdenum blue colour reagent. The intensity of the resulting blue coloration is measured in a colorimeter and is proportional to the phosphate content.

Craddock *et al* emphasize the value of the phosphate method as both a locational and interpretative tool, especially in combination with magnetic susceptibility, air photography and fieldwalking. We have already seen in the last chapter that phosphate measurements on topsoil were able to detect an underlying site covered by 0.5 m (1.6 ft) of alluvial flood clay. Proudfoot warns that variations in underlying site geology are likely to affect natural phosphate levels. The recognition of human activity is therefore much more straightforward on a site of uniform geology. Sandy and peaty soils tend to be the least satisfactory because of the loss of phosphates by drainage, although there is a compensatory tendency for plants to return them to the surface.

Eidt (1977) takes phosphate analysis a stage further by dividing the phosphate into three fractions which he analyses separately and defines as follows: (1) easily extractable; (2) tightly bound or occluded; (3) occluded calcium phosphate and apatite. He found the approach useful for distinguishing naturally occurring phosphates from those associated with human occupation, which also showed a characteristic balance between these fractions, irrespective of soil type, and characteristic fractionation patterns for different types of land use. For instance, he found a close match between a modern soil used for growing mixed vegetables and a prehistoric soil which was probably used for the same purpose.

Testing possible sites of eighteenth-century French forts on the Mississippi in Illinois, Weymouth and Woods (1984) used magnetometry and a suite of chemical tests – phosphate, calcium, pH (acidity/alkalinity), magnesium, potassium, iron, zinc and copper. The first three were the most effective, giving anomalies closely echoing those of the magnetometer. This proved to be responding to the foundation trenches of walls and to internal buildings. It is known that French construction at that time made use of lime mortar mixed with straw or animal hair, and the walls were built of limestone. The chemical anomalies showed high phosphate, calcium and pH (alkaline), which strongly supported the hypothesis that this type of wall construction was used.

Bintliff (publication forthcoming) reports work on sites in Greece, where enhanced concentrations of copper and lead in the soil are indicators of the spread of human activity.

Acoustic reflection

Thinking of the high effectiveness of seaborne sonar, people often ask if this method can be used for archaeological prospecting. In fact, water has enormous advantages in its uniformity, contrast in properties with its bed material, and the ease with which acoustic transmitters and receivers can be coupled to it. Soil may be shallow, but is notoriously complex in comparison. Weymouth (1986) reports some success with the method in the Middle East and Egypt, and promising experimental work has been done in Japan by Ozawa and Matsuda (1979). The latter used an assembly of a hammer-type vibrator surrounded by four geophone sensors 0.5 m (1.6 ft) away whose combined signals were computer analysed to produce a mean response compensated for inhomogeneities of the ground. Initial tests were successful in locating a stone coffin and indicating its depth. The soil was quite uniform but, even so, reflections from odd stones buried in it were rather obtrusive. The technique clearly has some relationship with radar, although the frequencies used are very much lower, and also with bosing.

An allied technique is seismic refraction. The depth of an upper layer of soil can be estimated by measuring the time of wave travel between a vibrator – a hammer striking a plate on the ground will do – and a series of geophones set in the surface at some distance. Using comparisons between the time taken for the direct shock wave to travel to the geophones with the time taken by a wave which passes into and along the lower layer, the depth of this layer can be estimated. Substantial archaeological features can be expected to affect this pattern of response, and the method was found by the Lerici Foundation to be marginally responsive to Etruscan tombs. A number of experiments cited by Weymouth give a rather gloomy picture, but an appropriate and successful use of the technique was for establishing the shoreline adjacent to Fort Frontenac, another French fort in an urban situation in Canada (Woods and Krentz, 1984).

Metal detectors

Less expensive metal detectors normally consist of a single or double coil which forms part of a tuned oscillator circuit. The presence of metal upsets this condition and produces an out-of-balance signal. The range of these instruments to small metal objects is only a few centimetres or inches, and better results can be obtained from the more expensive detectors based on the pulsed induction principle (Chapter 4). Flat laminar objects such as coins are readily detected because they are well shaped for the generation of eddy currents upon which the response depends. Modern instruments are normally designed to discriminate between ferrous and non-ferrous metals.

The extraction of metal objects for their own sake has little place in archaeology and has led to the notorious damaging of sites by treasure hunters, who not only remove the value of the metal object as a source of evidence but often also destroy the stratigraphy of its context. The location of metal is only part of the totality of information sought in excavation.

Under archaeological control, detectors can be useful in site survey for locating relevant artefacts, especially in disturbed topsoil. They have on occasion been used to clear sites of iron interference as a preliminary to magnetometer survey, but modern recording techniques are so well able to discriminate such interference that this is no longer necessary.

During excavation, metal detectors are useful for giving advance warning of metal artefacts, especially coins which can be difficult to see yet often provide crucial dating evidence – detectors can be as useful on the spoil heap as on the actual site! They are more likely to be helpful on an excavation which has to be done quickly under rescue conditions than on one where there is time for meticulous attention to detail.

Thermal sensing

The flux of heat in and out of the soil is affected by the thermal capacity and conductivity of its components, so that it is possible for buried archaeological features to affect the temperature at the surface and below. Perisset and Tabbagh (1981) have established the basic parameters of the phenomenon in temperate conditions, and the rather exacting requirements for detection of temperature variations which rarely exceed 1°C. The sensors are airborne scanning radiometers working in the near infra-red part of the spectrum (typically 10.5–12.5 μm), which produce a thermal image of the ground beneath.

Best conditions obtain about four days after a change in the temperature of the air mass between overall high and low, in either direction, after which the effect diminishes. These conditions occur more reliably in the stability of a continental climate than in the Atlantic climate of a country such as Britain. The authors show convincing and useful pictures of radiometric scans, which in fact can respond to differential heating by the sun of features with slight relief as well as the conductivity effect. All this work is carried out over bare soil, to avoid the masking effects of vegetation.

The quite familiar pictures produced by false colour infra-red film, on the other hand, specifically exploit the vegetation effects, responding to the thermal emissivity of crops which is highly sensitive to their condition and health, in turn strongly affected by the state of the underlying soil.

These airborne techniques hardly fall within the remit of this book. Measurements can also be made at ground level, preferably with thermistor probes placed in prepared holes some centimetres or inches below the surface to avoid

emissivity problems. Unfortunately such measurements are very slow for practical application. Experiments by Noël show that anomalies arise either by interaction of shallow features with the strong daily heat flux, or by interaction of deeper objects with longer-term changes in surface heat flow. The striking indications of buried remains, such as walls, which are sometimes revealed by partial melting of light snow cover, are the result of self-optimizing conditions that are difficult to emulate instrumentally: the snow is ready to melt at any time that the heat flux from the ground is suitable, when all 'readings' are taken simultaneously. Also, once melted, its change of state is irreversible. (See Scoller *et al*, 1990.)

Induced polarization

This method is well established in 'big' geophysics, especially for detecting ore bodies. The possibility of using it for archaeological detection was studied by Aspinall and Lynam (1970). When a direct current is passed through the ground, voltages can be induced at the interfaces between some constituent materials as a result of complex electrochemical effects, for instance between clays and ionic conductors, of which, as we have seen in Chapter 2, groundwater is one. Measurements were made with a four-probe array of non-polarizing electrodes by passing a DC energizing pulse through the ground, and then integrating the induced voltage over a period of time.

The method gave a rather better response than resistivity to a humus-rich ditch cut into Millstone Grit in a moorland setting, and also over a dump of highly polarizable coal shale; but it has not been further exploited. One deterrent is the need for non-polarizing electrodes; these are rather complex and vulnerable, having a porous structure and containing copper sulphate solution which itself acts effectively as the electrode. Although a relatively robust version of these was developed, rather slow operation and increasing refinement of resistivity have been against the technique in archaeology. The paper describing it is now better remembered for its introduction to the twin electrode probe configuration.

Dowsing

This ancient method, apparently capable of locating water, has been extended with reckless profligacy to the detection of absolutely anything the customer requires. The technique hardly needs reiteration: the detector is a forked stick held between two hands, or a pair of bent wires held horizontally side by side, both forms of detector actually serving to amplify some sort of muscular spasm or contraction in the user. The bent wire detector can be obtained in elaborate versions, with rods that rotate in bearings in handles.

One explanation for the dowsing response is that it is caused by the body responding to magnetic changes, either as a reflex action or by way of the subconscious brain. In a review article, Williamson (1987) has developed the magnetic theme. We have already encountered in this book the probable role of bacteria in enhancing the magnetic susceptibility of soils. It has been shown that some bacteria are actually steered by the torque of the Earth's field acting on minute chains of magnetite crystals within them. Similar magnetic material has been found in bones at the front of the skull in birds, which clearly use the Earth's field for navigation, and vertebrates, including humans (Thompson and Oldfield, 1986). The yellowfin tuna seems able to detect fields down to 1 nT. Baker (1989) claims that humans also have the ability to navigate by their own inherent magnetism, although the faculty does not seem very reliable. Williamson suggests that dowsers may find water by detecting magnetic anomalies associated with rock faults through which it runs, but an experiment by Aitken showed that a dowser was incapable of detecting Roman pottery kilns, which were relatively strongly magnetic.

Some respectable archaeologists swear by dowsing, yet there has been an almost total dearth of proper plans based on such surveys. This has been partly remedied by a recent book, *Dowsing and Church Archaeology*, by Bailey, Cambridge and Briggs (1988), in which the theme is developed from a standpoint of proper scientific scepticism. Experiments were carried out in controlled conditions: all surveys were planned and a few tested by excavation which produced some with impressive correlations, some less so.

Many controlled tests of dowsing have proved totally negative, and many results remain speculative and untested, or scientifically incredible, for instance the 'imprint' effect, by which the dowser appears to be able to detect structures no longer present. But the growth of knowledge has overturned scientific orthodoxy more than once, and it is seemly to keep an open mind.

Chapter 6
Choice of method:
choice of site

The previous chapters may well have left a rather blurred impression of a number of techniques with different capabilities, and a feeling of uncertainty about which to actually use when confronted with a site to survey. This chapter provides some guidance, as well as a brief indication of methods for judging the archaeological potential of a landscape.

Choice of method

Fig. 93 is an attempt to summarize the kind of response to be expected from each of the standard techniques to typical features on a straightforward British lowland site on fairly level ground with even soil cover – the kind of site that might show as typical crop marks from the air. I must acknowledge the pioneering use of this type of presentation by Greene (1983).

A reasonable magnetic susceptibility contrast between natural and topsoil is assumed. At the top is a schematic diagram of the site as it originally was, with the subsurface features shown in cross-section. A and B are a medium-sized ditch and bank of the kind that would have surrounded a settlement containing the Iron Age round house, D, with a central clay surface hearth (shown black) and a dump of domestic waste, or midden, C, outside. The house has an outer ring of posts set in a circular foundation trench, and inner posts with individual post-holes. E is a bell-shaped storage pit and F a shallow pit or scoop. G is a fenced animal compound and H a small Roman building with substantial outer wall foundations and the lighter foundation of an inner wall. Finally, J is a pottery kiln of baked clay, together with its stoke-hole.

The second diagram shows these features as they appear now, after prolonged ploughing. The visible remains of the round house are reduced to the bottoms of the foundation trench and post-holes, while the remains of the hearth have been

ploughed into the topsoil. The cattle pen is also only represented by post-holes. The Roman building has been robbed for its material but the lower parts of the foundations remain, with a scatter of smaller fragments in the soil. The superstructure of the kiln has collapsed into the interior. Relatively high concentrations of occupation soil affected by burning are shown by stippling.

The responses of the different techniques are shown as idealized graphs below the two cross-sections. The figure as a whole is designed as a quick reference in its own right, but the following comments provide some additional information.

Resistivity This is the most complicated technique to show, because of its dependence on climatic conditions. To start with, the time is assumed to be the latter part of the year when a net moisture deficit gives generally optimum conditions. The lines represent behaviour over extreme soil types, the continuous line chalk and the dotted line gravel. An approximation to the chalk response can be expected over clay, loam and other close-textured naturals, while well-drained sands will behave similarly to gravel.

The ditch A contains a relatively loose soil filling on chalk, and gives a positive anomaly. On gravel, this is likely to be negative because the gravel is well drained, and also because the soils that develop on gravel often contain alluvium and so are relatively silty and close-textured. The bank has been levelled but its effect may still be faintly detectable from its remnants ploughed into the topsoil, which may also be slightly shallower where the bank has protected the natural from erosion. This may give a slightly lower reading on chalk, but higher on gravel.

The midden heap, C, is not detected, nor are the post-holes of the house, D, because they have too small a volume but slight responses can be expected from the foundation trench. The pit, E, is shown as giving a small negative response on chalk because its filling is likely to be original, organic and close-textured. This will be accentuated on gravel. Pit F is more exposed to evapotranspiration and ploughsoil filling, and gives a response similar to, but weaker than, the ditch.

The building, H, produces a series of clear peaks over the walls, superimposed on a general rise due to the rubble spread. These responses will be suppressed on gravel, which has a comparably high resistivity, and on which the detectability of the remains depends chiefly on their being at least partly buried in soil above the gravel. It is noticeable from the air (for example, Wilson (1982), Fig. 28) that many buildings, as well as more primitive sites, are surrounded by a dark penumbra of relatively damp soil, often due to clay from the ruined walls and sometimes probably organic detritus. These are likely to give a generally lowered resistivity in the region of a building, and of settlement generally. Some such response might even be obtained from the cattle compound, G. The kiln and stoke-hole, J, will behave similarly to the building and ditch respectively.

When the moisture content of the soil is at field capacity, usually during the first half of the year in Britain, these responses will lose their clarity or even change their sign. The changes are discussed in the following paragraph, but not illustrated.

We have seen in Chapter 2 that in such conditions ditches on chalk can become virtually undetectable, or negative anomalies. A medium-sized ditch like A will probably become slightly negative. On gravel, increasing short-circuiting by the superincumbent soil will tend to more than balance the lower resistivity of the ditch filling, causing a reduction of the negative anomaly. The foundation trench of D will probably be undetectable. The response to pit E will probably diminish on chalk but not change much on gravel, where these factors will balance out. F is likely to disappear on chalk, and weaken on gravel. The response to the building, H, will be diminished on both, depending on the porosity of the construction material, but it should always remain detectable, as will the kiln, J. The stoke-hole of J is likely to become a weak negative anomaly on chalk, and to diminish on gravel.

Magnetometry The response to ditch A will tend to be reduced by the primary silting and the falling back of bank material from B, both of which will contain a large proportion of the less magnetic lower layers. Magnetic contrast will also be low if a clay bank has silted back into a clay-cut ditch. However, in both cases, the detectability of the ditch will probably be helped by the spread of magnetically enhanced soil from the house, D.

The foundation trench of D should be just detected, the magnetic enhancement of the component of its filling derived from the house compensating for the small cross-section. There will be a strong response to pit E because the magnetometer will see the signal from the whole of its bulk, which will include much magnetically enhanced occupation soil. The signal from pit F will be much weaker because it approaches the laminar shape to which magnetometers are insensitive, like the hearth in house D. The wall foundations of the building, H, will give weak negative signals in contrast with the relatively magnetic soil unless, for instance, the walls are built of brick, or igneous stone brought on to the site, in which case they will show as chaotic positive and negative anomalies. The response to the kiln, J, will be particularly strong, and to its stoke-hole substantial.

Magnetic susceptibility A weak positive signal may be produced by material brought up by worms or ploughing from ditch A, while bank B may give rise to a small negative signal due to dilution of the topsoil by bedrock material from it. There will be a slight response to the ploughed-in remains of the midden, C, because it is bound to have included some burnt material from the house, D. The site of the central fire of D is likely to show clearly, and the whole area of the house will be enhanced by material spread from this fire. There will be some response to the spread of occupation soil in the area of E and F, which themselves may be just resolved as separate anomalies caused by the scattering of some of their contents into the topsoil. Such pits (and ditch A) may be directly detectable by more deeply penetrating electromagnetic instruments, but with limitations that will be discussed in the summary below.

Enhancement may be expected from material spread from the domestic fires of building H, but this will be partly counterbalanced by dilution of the soil by the rubble. Substantial enhancement will also be caused by the burning associated with the kiln, J.

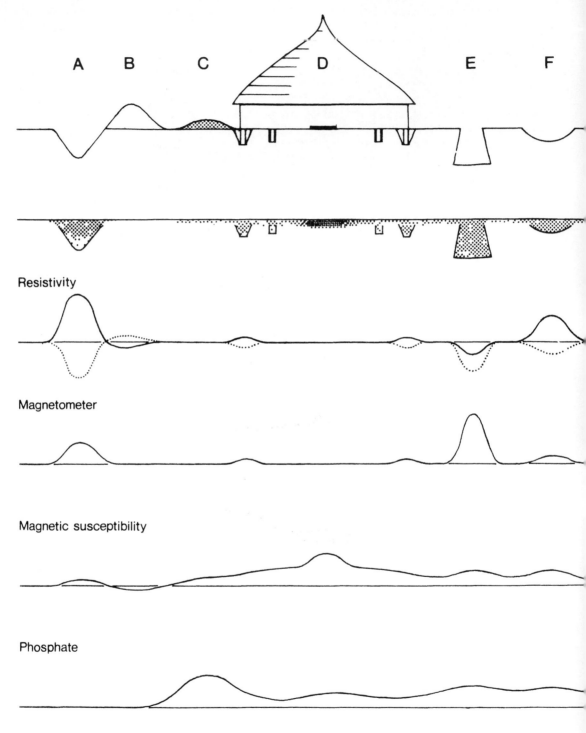

Fig. 93. (Top) Various ancient structures before and after conversion to an archaeological site. (Below) Responses of different prospecting techniques.

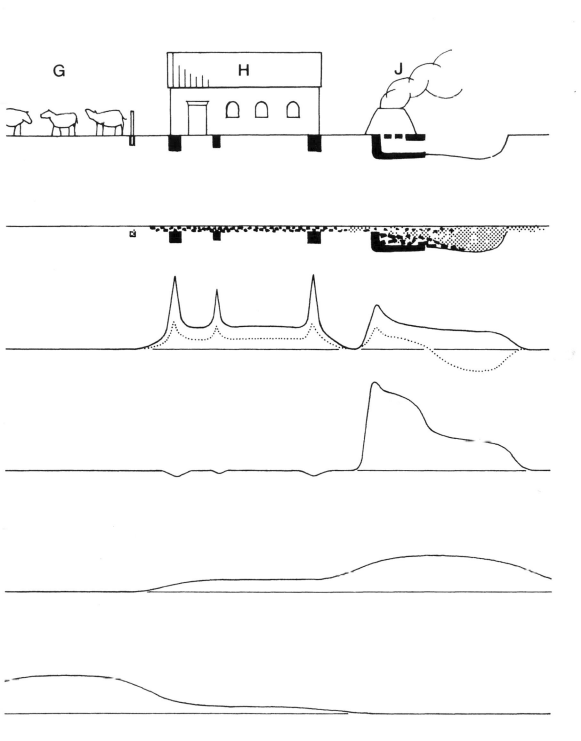

Phosphate This will respond most clearly to the midden deposit, C, and the cattle pen, G. There is also likely to be some response to the detritus of domestic activity in the area of D, E and F, especially around the pits, and in building H.

Summary The striking thing about the various techniques is the degree to which the information they produce is complementary (Fig. 94), so that the fullest non-invasive information about a site can only be obtained by using all of them. However, few real sites are as varied as this hypothetical one. It would generally be sensible and economical to use magnetometry for an Iron Age site and resistivity for stone buildings, with perhaps a preliminary broad survey by magnetic susceptibility or phosphate, or both. Speed and convenience tend to favour the magnetic methods, and intensive magnetic susceptibility survey must always be kept in mind for obtaining the maximum information on sites where a large amount of surface activity is likely to have taken place when they were in use.

A limitation sometimes encountered with resistivity is exemplified by pit E. Although substantial, this can give a smaller anomaly than the far less massive pit F, partly because of the likely moisture balance mentioned above, but mainly because of its narrow top, which has a masking effect.

Magnetic susceptibility measurements can be similarly skewed by their shallow depth of detection. Even the deeper-penetrating instruments will tend to respond mainly to the tops of features, so that a shallow pit can look like a deep one and, worse still, a thin occupation layer may be indistinguishable from the pits. The effect of this lack of discrimination is to give rather diffuse pictures of occupation areas. Magnetometers suffer from the opposite problem of being insensitive to the thin layers, but they are especially effective for distinguishing the relative sizes of pits because they respond to their total magnetic bulk. Thus, a dual magnetometer and susceptibility survey can be very informative.

Many sites, of course, are not as ideal as this one. Erosion may be severe, leaving little to detect except perhaps primary silts having little magnetic contrast with the bedrock, while the soil may have drifted away downhill leaving magnetic susceptibility instruments nothing to detect. The primary silts will tend to develop high resistivities

in the dry period, so that this might be the most effective method. Conversely, a site where soil has accumulated during occupation will probably have many features cut or encapsulated in the soil which are difficult to see. Here susceptibility may be a valuable aid during excavation: surveys at successive levels of clearance may well be more sensitive than the eye.

Resistivity can also be very sensitive over such deep deposits. The example shown in Fig. 95 is of a Saxon settlement which is being excavated by the Wraysbury Archaeological Group. The site is on sand, and about 1 m (3.3 ft) of dark soil contains a complex sequence of occupation. Very slightly increased readings revealed distinctive rectangular shapes with a common orientation and an appropriate size for Saxon huts. It was difficult to recognize anything of these in the trenches, although one or two post-holes were found in appropriate positions. The survey was presumably reacting to invisible earth floor levels whose water content was reduced by compaction even though most of the survey was made in spring, a poor time for resistivity contrast on many sites. A discontinuity between the left and right parts of the survey is due to a change in conditions during an interval of time; it illustrates the constantly changing delicate balance of the moisture regime on which the technique depends.

Chapter 2 has already demonstrated the very high sensitivity of resistivity in conditions of net water deficit, for instance, its clear response to the berm of the Hog's Back barrow. The Roman villa at Compton was also mentioned, where a 0.5 m (1.6 ft) twin electrode survey was made on 13 September 1987. Narrow anomalies in chalk-based ploughsoil rose to such values as 511 ohms against a background level of about 347 ohms – almost a 50 per cent increase. These were interpreted as building walls, but when this anomaly was tested by excavation it proved to be a slot about 60 cm (24 in.) wide cutting about 25 cm (10 in.) into the natural chalk from a depth of 57 cm (22 in.), apparently a gully along the uphill side of a slight terrace. This could have been a cultivation terrace associated with the Roman villa. So impressive was the effect of this slight feature that one is tempted to recommend making resistivity surveys in search of buildings during non-optimum periods of net water gain, to avoid being misled. The ideal, if time is on one's side, would probably be to make surveys in both conditions,

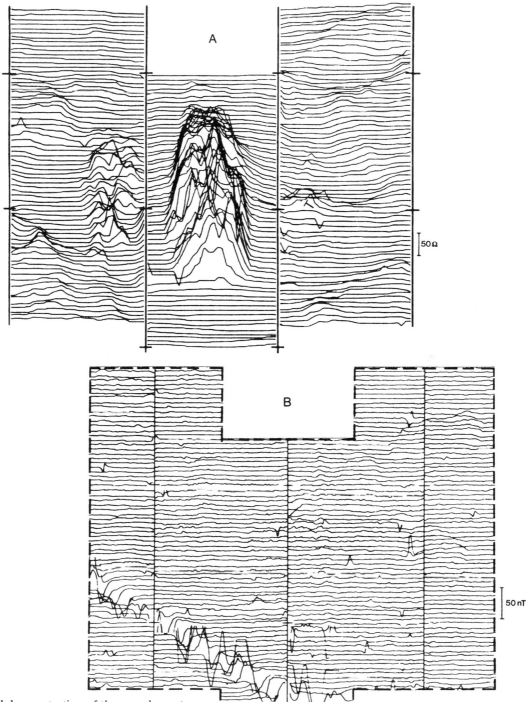

Fig. 94. A vivid demonstration of the complementary response of resistivity and magnetometry. (A) Resistivity survey of a stone-built long barrow in the Cotswolds. (B) Fluxgate gradiometer survey of the same area shows little response, except to a modern pipeline which is invisible to resistivity.

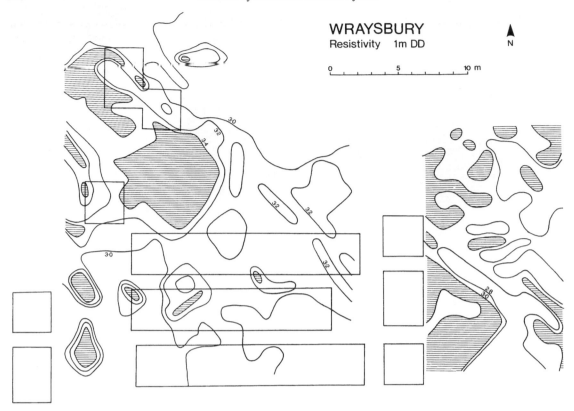

Fig. 95. A very sensitive resistivity survey of a deep occupation deposit over sand natural, showing outlines of probable buildings hardly detectable in excavation. Although the anomalies are very slight, low background 'noise' allows the effective use of contours, with the assistance of shading for the highest levels.

to plan the building and its more subtle surrounding features.

Resistivity in optimum conditions is probably the best, or at least the most economical, option for finding soil-filled features where the soil is weakly magnetic and there has been little human occupation to cause enhancement. An example was the Ancient Monuments Laboratory survey of Avebury, where the detection of stone-holes was a major objective. Another contender is ground-penetrating radar, although it would be slow and costly to achieve comparable ground coverage by this method.

When harsh economic reality is breathing down one's neck, demanding speed and value for money, the choice seems to converge on the fluxgate gradiometer. Advising recently on the best approach to archaeological prospecting along a pipeline route, I found myself recommending in every case the use of detailed fluxgate survey or scanning, or both. Only for one particularly sensitive place did I tentatively suggest a susceptibility survey as a possible source of supplementary information. Even where magnetic detection is not ideal, the superbly sensitive plotting techniques now available mean that there is a good chance of some indication of any archaeological pattern being revealed, even in adverse and unsuitable conditions.

Geomorphology and archaeological potential

A Greek geologist recently complained to me that archaeologists had a rather undue propensity for seeking archaeological sites on coastal land which did not even exist at the time in question. Such problems are hardly likely to arise when a specific known site is being examined, but are worth considering if, for instance, a scan is being planned for a new roadway route. Problems in low-lying areas can range from the total absence of land in ancient times to masking by heavy blankets of alluvium or peat. Expert advice should be sought. For Britain and the Netherlands, the book *Archaeology and Coastal Change* (Thompson, 1980) is a valuable source of information.

Preliminary geological investigation can provide crucial information about the archaeological potential and limitations of an area, and thus an aid to the allocation of resources. A good example is provided by the systematic work undertaken in advance of the building of a lock and dam on the Monongahela River in Pennsylvania (Hamel and Jacobson, 1988). Earlier archaeological work had shown that there were likely to be Native American dwelling sites in the area, but was inconclusive regarding the depth to which the activity might have penetrated. Comprehensive geological and pedological investigation showed that the river terrace in question had been deposited around 80,000–60,000 BP, so that deep archaeological stratigraphy would be absent and most of the evidence would be confined to the 30 cm (1 ft) of overlying soil. The validity of this conclusion was confirmed by excavation.

Information for this kind of study in Britain is available from the NERC British Geological Survey Geosciences Data Centre, Keyworth, Nottingham.

Chapter 7
Interpretation and presentation

When measurements have been completed in the field, only half the battle is over. Just as important are the methods for extracting the maximum of archaeological information from them and their presentation in a way that is most comprehensible and acceptable to the archaeologist, and indeed to general readers of their reports and books. The objective is nothing to do with numbers, which are the raw material, but simply to produce the clearest possible archaeological picture.

We have already seen that different types of feature give rise to different characteristic instrument responses, which are readily seen in the kind of profile produced by graph or trace plots. Individual profiles of this type can be very telling and were used particularly frequently in the early days of resistivity. However, the use of horizontal grids of readings not only adds credibility to features by showing them in plan, but also provides an important reinforcement effect for responses that are too slight to be reliably interpretable in individual graphs. Much of the presentation battle has consisted of efforts to show surveys in plan without sacrificing the vertical resolution of profiles.

An important constraint on plotting techniques has been the need for reproduction in reports, journals and books. For the sake of economy, the attractions of colour have been largely resisted, although this is becoming increasingly accessible and reasonably priced. Black-and-white presentation must be reproducible by photocopying and printing processes, so that a compromise has to be achieved between the fineness of treatment needed for the subtlest presentation and the loss of this quality that may occur in reproduction.

Trace plots and manual interpretation

Before convenient digitizing equipment was available, the use of stacked trace plots was the easiest and most informative way of recording the output of continuously reading instruments, and was used with fluxgate gradiometers by the Ancient Monuments Laboratory for 15 years. Some fine examples are shown in Fig. 97; others are published, for instance, in Hinchliffe (1986) and Lambrick (1988). Its high resolution greatly facilitated interpretation, including the confidence with which we could recognize non-archaeological interference, especially from modern iron (Fig. 96). Its value was so much appreciated that even modern digitized data are still routinely plotted out in this way to assist with interpretation (Fig. 105) and resistivity surveys are similarly treated. But there are drawbacks. From the plan point of view, all peaks representing features are displaced from the traverse to which they belong, so that they do not appear in their actual positions and the plan is distorted; the more substantial the response, the worse is the effect, with added confusion if a number of large peaks overlie one another. Secondly, as we have seen in Chapter 3, linear features parallel to the traces can be poorly represented, perhaps by the inconspicuous displacement of a single trace (Fig. 58). Thirdly, clarity can be lost because of the continuity of traverses between anomalies, especially when large surveys have to be presented at a reduced scale. For these reasons, trace plots are used for final presentation quite rarely, or alongside plots prepared in a different way.

On a few occasions in the early days, as also mentioned in Chapter 3, trace plots were digitized by means of a manually operated trace follower so that they could then be converted to dot density plans. In addition, the surveys had to be edited first to disentangle overlapping traces, which were picked out in colour. This was tedious

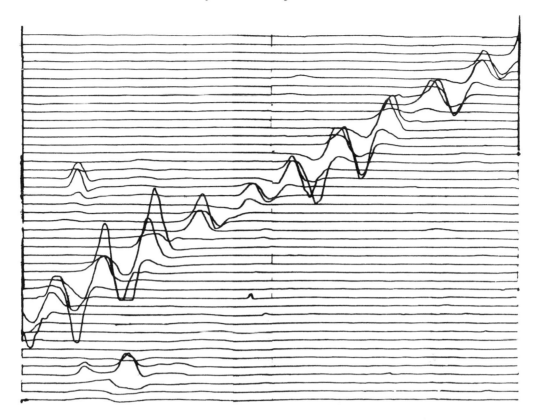

Fig. 96. Hazards of trace interpretation. There are kilns in these plots, but the row of alternating high and low peaks is a pipeline leading to a cattle trough. The regular north and south poles show that the pipes must have been about 21 ft (6.4 m) long and made on the same plant, so that they were magnetized by the Earth's field in the same direction. 50nT/line interval.

and expensive – and uncomfortably inappropriate to the ethos of the computer age. Modern digitizers are more efficient, but they do not circumvent the major problem of overlapping traces.

The art of producing manual interpretative plots became highly developed in the Ancient Monuments Laboratory. They were prepared on overlays on the trace montages, and took various forms. At small scales, anomalies were blacked in; on larger scales, they were outlined to produce what we called 'sausage plots'. A variation with a semblance of sophistication was the use of dry transfer tones which gave a dot density, or half-tone, effect and the opportunity to show some indication of different anomaly strengths. The element of subjectivity inherent in manual interpretation, anathema to the purist, can be beneficial so long as preconceptions can be kept at bay. The human brain, after all, is a very flexible computer, able to judge more reliably than an electronic computer what looks like modern interference or instrument problems; and it can bear in mind the geology and reject the type of anomalies it may produce, for instance, the igneous dykes of Orkney which can so nearly mimic buried ditches.

A good example of manual interpretation using transfer tones is the survey on chalk at Barton Stacey, Hampshire (Fig. 98), on the line of the British Gas Southern Feeder pipeline (Catherall *et al*, 1984). The directly recorded traces are rather subdued and confusing, but the broad interpretation is quite clear. A rhombus-shaped ditched enclosure seems to have its main entrance at its eastern corner, from which a wide ditched track, possibly a drove road, runs away. The northern side of this seems to be picked up again on the north-west side of the survey, and may be a boundary against which the enclosure

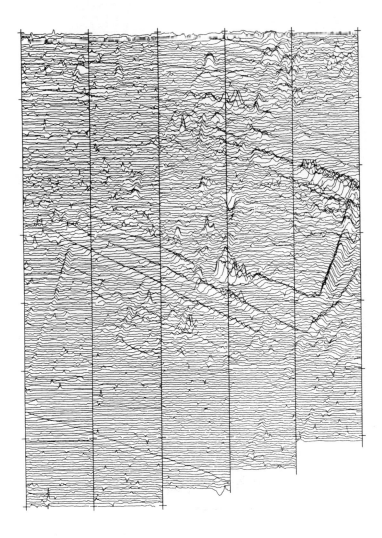

Fig. 97. The apotheosis of direct fluxgate gradiometer trace recording. (Above and top right) A 5.43 ha (13.4 acre) survey of a site at Wharram Le Street, Yorkshire. A rectanglar area of noisy readings along the north side of the prominent central square ditched enclosure was interpreted as a Roman villa building. A diagonal line of 1 m (3.3 ft) square test holes at 20 m (66 ft) intervals produced valuable complementary evidence of the type and condition of remains (Rahtz *et al*, 1986), revealing occupation from prehistoric to Roman times, and confirming the building. There is also a weakly defined circle near the centre of the enclosure. (Bottom right) Survey on Thames gravel at Radley, Oxon, showing very clear Bronze Age ring ditches. Sites on the upper Thames often show well because their material is partly derived from the Jurassic geology of the Cotswolds. Both surveys: 30 m (100 ft) strips; sensitivity 7.5 nT/line interval.

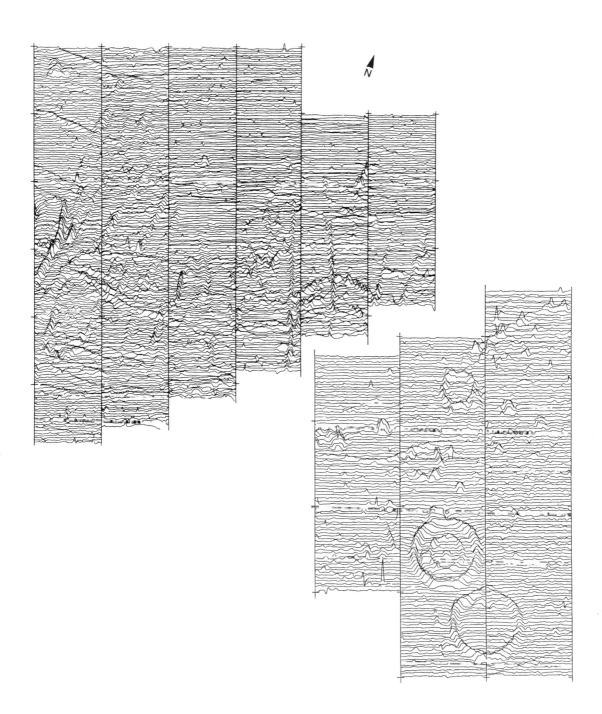

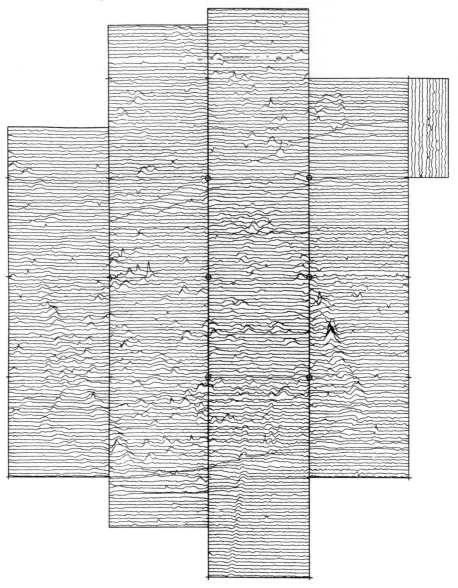

was built. This possibility is supported by the narrower double-ditched road coming up from the south-west, which changes to a simple hollow way after crossing the putative boundary. This suggests that the road needed to be clearly marked, and perhaps fenced, in the heavily occupied area south of the boundary but that it entered more open ground to the north, perhaps pasture, where its definition was unimportant. In fact, the survey showed that the enclosure must have been out of use when the road was established; therefore the change of character of the road suggests that the boundary continued to be significant after the decline of the enclosure.

The evidence for the relative dating of the road came from its intersection with the south-west ditch of the enclosure. The traces show the road ditch anomalies running through the more substantial enclosure ditch, which would have obliterated them if it had been cut later. The anomalies would probably also have been interrupted if the ditches had been cut only into a naturally accumulated organic filling, and it therefore seems likely that the ditch had been

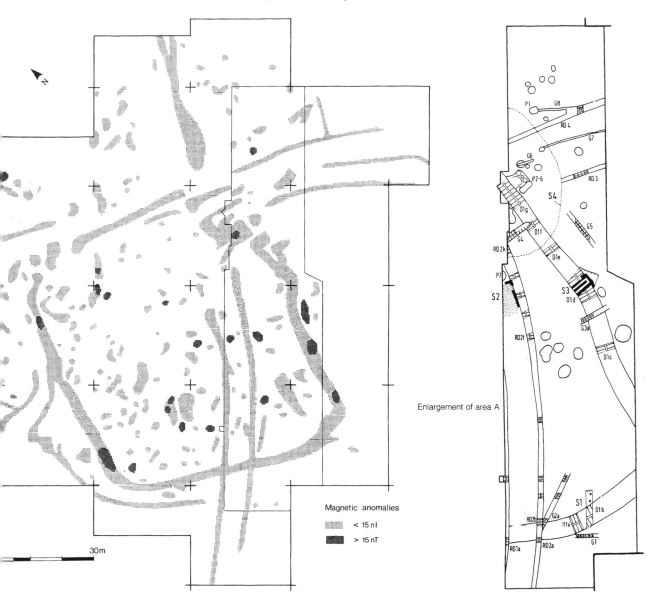

Fig. 98. The Barton Stacey project. (Opposite) Montage of survey traces (7.5 nT/line interval). (Above) Interpretation. (Right) Partial excavation.

Enlargement of area A

Magnetic anomalies

< 15 nT

> 15 nT

30m

slighted by the throwing down of the bank into it, producing a chalky upper fill with a lower magnetic susceptibility than the subsequent silting of the road ditches. The enclosure bank was not thrown down just to provide a causeway where the road passed through because its ditch anomaly is uniform along this side.

Two other aspects of the survey are worth mentioning. Near the north corner of the enclosure, the traces show a cluster of small spiky peaks probably representing genuine archaeological iron rather than the more common modern rubbish. They are associated with normal anomalies, and their subdued height indicates that they are quite deeply buried. They have normal polarity, without any reversals, so they must be soft iron detected by its susceptibility rather than having a strong magnetic field of its own – unless they are very compact small clay furnaces, perhaps the floors of iron furnaces. On the south-east side, a group of strong anomalies probably represents a

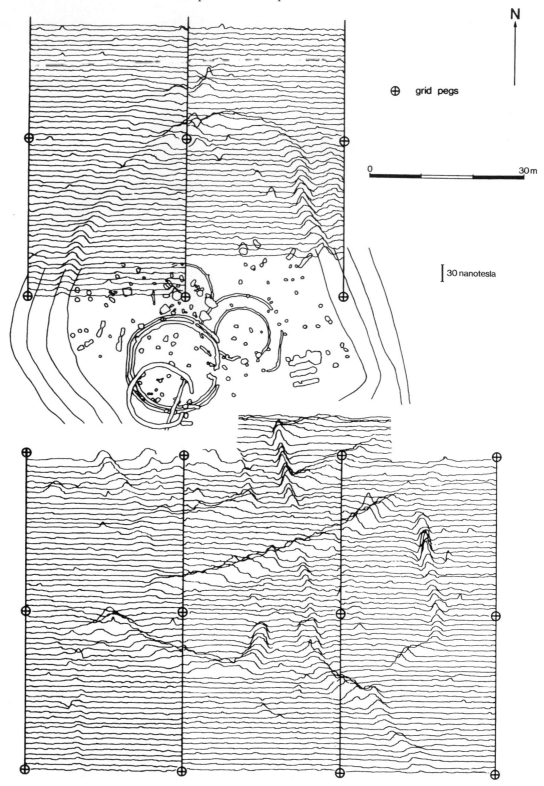

Fig. 99. The Groundwell Farm survey.

kiln or furnace built in the ditch, perhaps after it went out of use, and therefore contemporary with the late road. Many pits and minor ditches are scattered around the area.

It is significant that in the above analysis much more use has been made of the evidence from the traces than from the more visually acceptable plot, confirming the importance of retaining or generating traces for survey interpretation.

Such was the interpretation of the survey. Rapid trial excavations by the Andover Archaeological Society confirmed with several sections that the main ditch had been deliberately backfilled, and that the road ditches had been cut through the filling, probably during the second century AD. Curiously, a Roman corn-drier in the ditch at S3 had little effect on the magnetometer, but a substantial anomaly further southwest along the ditch was missed by the excavation.

A survey at Groundwell Farm, on the outskirts of Swindon (Fig. 99), was commissioned to complete the plan of a partially excavated site containing a sequence of Iron Age huts within a ditched enclosure (Gingell, 1982). The immediate bedrock is Coral Rag, part of the Jurassic series that can be relied upon to produce clear magnetic definition. The main ditches of the enclosure are beautifully revealed, but the limitations can also be seen. Although they were strongly magnetic, even the largest post-holes could not be reliably detected, even when the traverse interval was reduced to 0.5 m (1.6 ft); nor did the survey respond to the narrow foundation trench of a hut that it covered.

Again, something can be said about the development of the site from the survey evidence. It seems clear that it began as a 'banjo' enclosure, with a long, narrow entrance way on the south side, the ditches of which spread out to form 'antennae', which normally define fields grouped around the enclosure. In a later phase, the entrance has clearly been cut across with a blocking ditch, and the side ditches wholly or partially obliterated by a backfilling of bedrock rubble, probably excavated from this ditch.

A survey on a much larger scale was made at Lake Farm, just outside Wimborne Minster in Dorset. An extensive series of trace-recorded gradiometer surveys was successfully summarized with a manual plot, the outlines being blacked in because of the smallness of the anoma-

lies relative to the area of the survey (Fig. 100). The site occupies a tongue of slightly raised ground projecting northward into the floodplain of the River Stour, and was a Roman vexillation fort used as a supply base for the conquest of south-west Britain under the generalship of the future Emperor Vespasian. It had been recognized in meticulous excavations by Norman Field and Graham Webster along the route of a disused railway marked by the double broken line on the plan, now replaced by the Wimborne by-pass. The geological base of the area is complicated by the conjunction of different deposits, and consists of London Clay admixed with gravel and sand. This has a fairly high magnetic susceptibility, with the consequence that archaeological anomalies due to infilled features are generally rather slight because of lack of magnetic susceptibility contrast, although there is good enhancement where burning has occurred.

Initial small-scale work with individual traverses was of no value, archaeological anomalies being too small and background 'noise' too severe. Even the substantial south-western area first surveyed in detail showed very little, and it was obvious that a meaningful picture would only be established by fairly comprehensive survey cover. This itself was complicated by indications of more than one phase of activity, but the broad plan of the fort was satisfactorily established, including the lines of some internal roadways, recognizable from their side-ditches.

The detailed survey of the north and north-eastern area is worth discussing in detail (Fig. 101). The north-east corner of the fort has the characteristic Roman rounded form. South of it, there are signs of two ditches with a narrow negative anomaly between them. Along the north side, the outer ditch and the negative anomaly are the most noticeable. The negative anomaly presumably represents the ridge of undisturbed natural between the ditches. It is in fact very weak but is emphasized by the slight positive effect of the ditches on either side.

An arresting feature of the survey is the line of erratic strong positive anomalies that clearly run along behind the defences and must represent the kilns, hearths and furnaces required by Roman army regulations to be sited around the periphery of a fort to minimize nuisance and the danger of fire. Some of the internal arrangements of the fort show intermittently, and there are spectac-

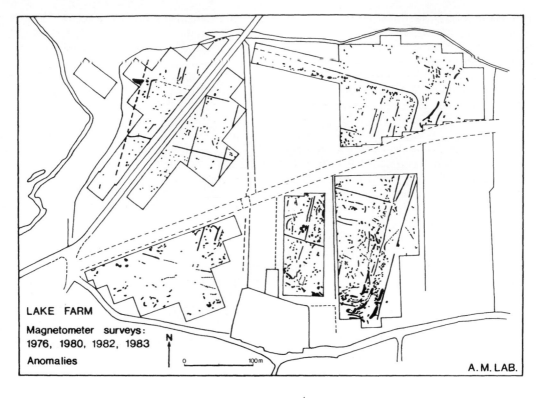

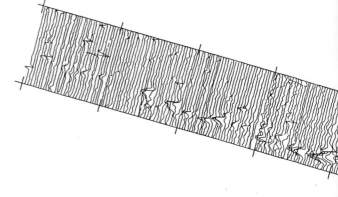

Fig. 101. Lake Farm Roman fort. North-east corner of fluxgate survey. Sensitivity 19 nT/line interval; strips 30 m (100 ft) wide. The extension following the defences has been stepped westwards to show two aspects of the anomalies at the overlap. See text for discussion.

Fig. 100. Lake Farm Roman fort. General interpretative plan.

ular positive and negative iron spikes representing pieces of magnetized iron that could be of any age.

Continuing westwards, the anomalies representing the defences become progressively weaker and finally disappear. The 'burning spikes' also diminish in strength. To complete the outline of the fort on this and the west side, it was necessary to rely on these as the only indication of the line of the defences. The explanation must be that this part of the fort has been covered by a deposit of alluvium from flooding of the River Stour.

One further point to make about the interpretation of the detailed plot of the north-east corner is that there are signs of a fair amount of extra-mural activity, which lacks the more organised planning of the fort. This could be part of a *vicus*, or military workshop activity away from the fort.

Contour or isograph plots

These are used quite rarely because they are effective in a rather limited number of cases. Where anomalies are clear-cut and coherent, giving steep gradients with closely parallel contours, their advantage is that they provide definite and easily comprehended outlines, but with weaker data they can wander about aimlessly and distractingly, outlining nothing in particular. On many plots, a mixture of the two is obtained. To achieve vertical sensitivity and good delineation of features, the contour interval needs to be close, but this tends to produce an increasingly fussy background so that selective contouring may be necessary. Although this sounds subjective, it is a fact that steeper gradients selected in this way are the most likely to be significant.

A major cause of 'wandering' contours is variation in background readings. This particularly affects resistivity surveys, and can often be improved by the use of filtering, discussed below. Another problem with contours is that it is difficult to convey which are the high or low values at

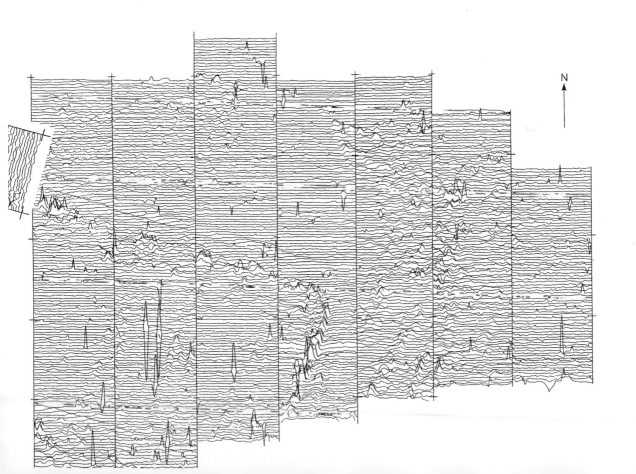

a glance. This has been partially solved by putting right-angled ticks, like small hachures, along the low-value side of the contours, but even this treatment requires quite close inspection to comprehend. Shading can be effective, but needs to be carefully used to avoid diminishing clarity. In the Wraysbury survey (Fig. 95), just one shading level – of the areas contained by the highest contour plotted – was sufficient to clarify the trends.

For computer contouring a good quality program is required, otherwise there is a tendency to produce curious angularities and lozenge shapes that are a further distraction.

Atkinson (1963) has recommended that the minimum significant contour interval should be

Fig. 102. Selective contouring. (A) Section trench and resistance contour plot of barrow quarry. (B) Computer contour plot used as basis for (A). Contours in ohms.

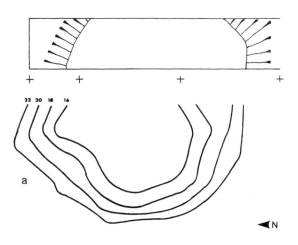

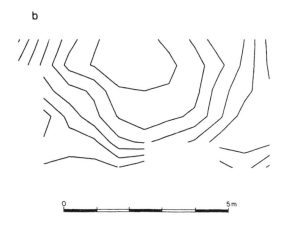

half the standard deviation of a set of readings. He made the points that if the interval is much smaller than this, the work of preparing the plot becomes excessive (he was writing in pre-computer times) and the difficulty of reading it may be increased; while a wider interval may miss significant anomalies. This remains a valuable basic starting point. An initial plot prepared on this basis, or simply by inspecting the data for obvious trends, will often show at what level the contours are defining significant gradients, and which contours (usually towards the higher or lower ends of the data span) are wandering aimlessly about the plan.

It is instructive to compare selective contouring with control data from excavation, which may even suggest a particular contour most suitable for defining the feature. At Sharpe Howes, on the Yorkshire Wolds chalk, trenching suggested that the material for a Bronze Age barrow was obtained from four quarry pits rather than the usual ditch (Fig. 102). Square array resistivity measurements plotted as selective contours strongly supported this. In the example shown, the 22 ohm contour probably gives the best representation of the quarry outline, although it is not perfect because the whole contour pattern is slightly shifted to the north in response to deepening topsoil on this side. An interesting detail is that the contours tend to turn inwards as they approach the trench, which has dried out the ground close to it.

Another example of this type of contouring is provided by the work at Druid Stoke, Bristol, on Carboniferous Limestone. Here a megalithic burial chamber stands in a domestic garden, and a twin electrode resistivity survey was used to test for any indication of an accompanying mound. Only a small group of contours was found to define a reasonably steep resistivity gradient, but this did form an outline containing the stones, suggesting a long mound with a plausible east-west orientation and a possible re-entrant forecourt in front of the stones (Fig. 103). The values selected for the contours were 85 and 90 ohms at the two ends but, in between, the gradient was defined by higher values of 100, 105 and 110 ohms – this was because the topsoil had been cut away for a garden terrace here, causing an overall lowering of the readings. Test trenches and foundation trenches revealed the very last remnants of a mound with limits broadly in agreement with

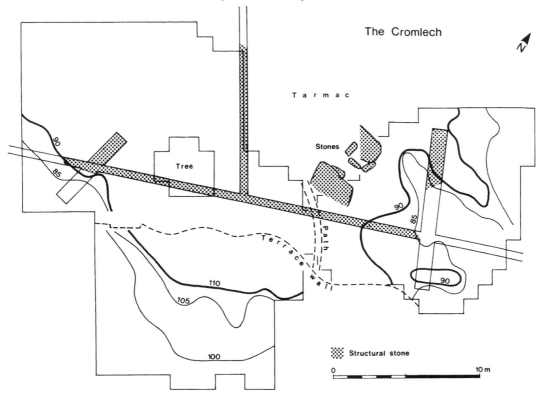

Fig. 103. The Druid Stoke survey.

the resistivity contours. The best definition of the mound was given by the the 90 ohm contour at the ends and the 110 ohms contour in the middle area.

It is perhaps appropriate to mention here, in the context of resistivity surveys, that Atkinson has advocated the conversion of all field resistance readings to resistivity for plotting, on the basis that this facilitates intercomparison of data. This sensible procedure has largely fallen by the wayside for two main reasons. Firstly, the twin electrode array is now the dominant survey method, and we have seen that it is difficult to determine resistivity values with this configuration. Secondly, the modern emphasis in plotting is on pictorial rather than numerical presentation, making such conversion irrelevant. However, if numerical values are used in plotting, as in the contour surveys discussed above, it is important either to convert to resistivities or to specify the type of electrode array and spacing, so that resistivities are determinable, or can be estimated, if required.

Dot density

This technique, first used by Scollar, has been until very recently the dominant method for producing 'archaeologist-friendly' plots of geophysical surveys (Figs. 104, 105). The objective is to create something approximating to a half-tone picture of buried remains, with some indication of the relative strength of anomalies.

Each reading is represented by a cell containing a group of randomized dots proportional in number to the difference between the reading and a predetermined baseline level. At the baseline, the cells are blank, and dot saturation – complete blackness – is arranged to occur at a level which will show up remains as clearly as possible without swamping detail among high readings. A dotting range from the mean (or close to it) to two standard deviations of the reading values above or below the mean is normally effective, but of course can be varied if the result is judged too heavy or too light. The dot distribution is slightly scattered beyond the boundaries of the cells, so that the pattern of the survey grid is suppressed as

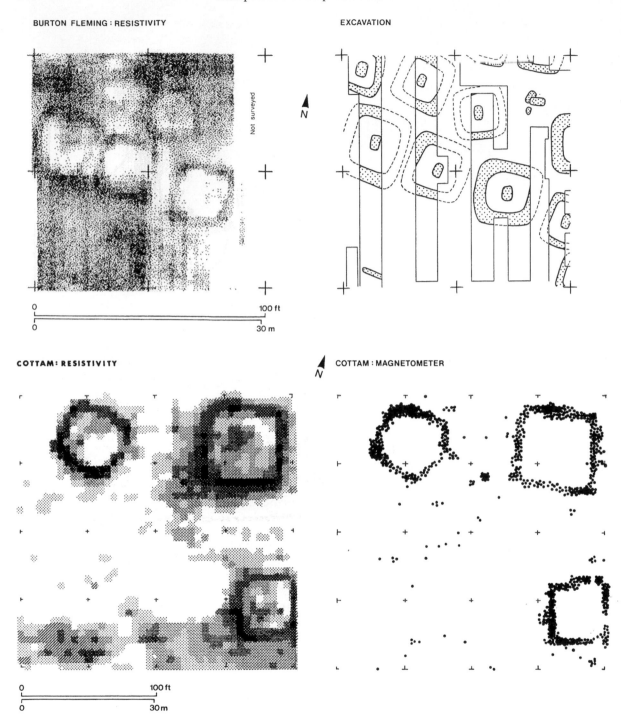

Fig. 104. Manually produced dot density plots of Iron Age square barrows surveyed by 0.76 m (2.5 ft) square array resistivity and proton magnetometer (see text). The barrows at Cottam (below) were much larger than those at Burton Fleming. Reading intervals: Burton Fleming, 2.5 ft (0.76 m); Cottam, 5 ft (1.52 m). Plotting limits: Burton Fleming, base level 18 ohms, saturating at 15 ohms. Cottam resistivity: base level 6 ohms, saturating at 2.5 ohms – a much larger percentage variation. Cottam magnetometer: base level 2 nT, saturating at 8 nT.

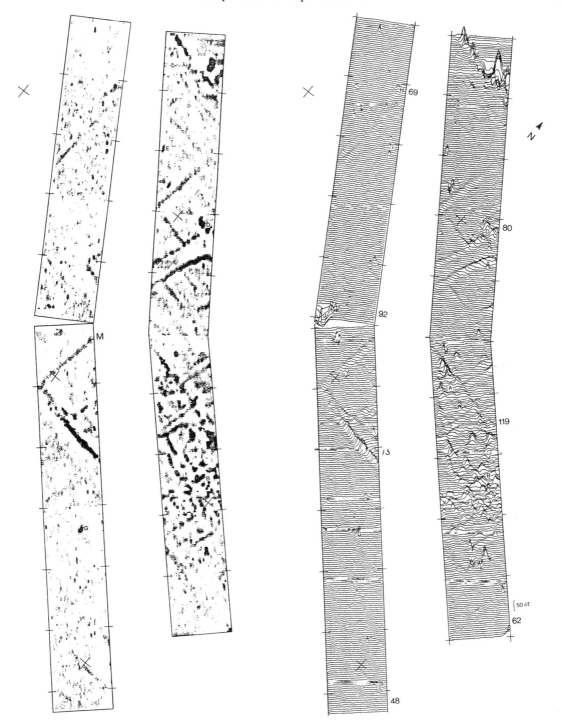

Fig. 105. Testing alternative routes for Latton by-pass, Wiltshire. Fluxgate gradiometer survey through a Romano-British settlement with dot density and trace presentation. Data slightly smoothed. The figures are topsoil magnetic susceptibility values in units of 1×10^{-8} SI/kg. The strips are 30 m (100 ft) wide.

much as possible, to avoid distracting the eye from that of the underlying archaeology. Increasing the dot density according to a power law will often add to the clarity of a plot.

To avoid undue cluttering of the dot density picture, and to present anomalies clearly, the art is to choose a base level that leaves plenty of clear background, without the weaker anomalies being lost beneath it – a balance which can be difficult to achieve to a nicety. It is normally necessary to produce separate plots for displaying high and low readings. Thus a magnetometer survey showing building walls as slight negative anomalies is often most clearly plotted with increased dot density in the negative direction, with a separate plot to show any of the more normal positive anomalies present. Alternatively the positive and negative anomalies can be shown together by choosing a very low cut-off level. Walls show in white against an overall high-density background in which positive anomalies will probably be visible but indistinct.

Interpretation is greatly helped if survey data are displayed in both plan and profile by the use of dot density alongside trace plotting. Fig. 105 is a fine example of this produced by Alister Bartlett and Basil Turton, showing part of a survey to test the archaeological impact of two alternative routes for a by-pass at Latton in Wiltshire through an area of complex Romano-British settlement. Unlike those in Fig. 96, the traces are constructed from the digitized signals also used for the dot density.

The main part of the settlement is clearly shown in the more easterly strips as anomalies of 10–40 nT. The figures alongside the trace strips are mass-specific magnetic susceptibility values for samples taken from the strip centre lines, and also closely reflect the more densely occupied areas. An important point illustrated by this survey is that dot density sometimes appears disappointingly weak in comparison with traces, for instance the ditch marked M, even if, as here, the dotting base level has been set low to pick up as much detail as possible. The traces appear stronger because of the emphasis provided by the slight negative anomalies that accompany the positives. Judicious filtering can be applied to bring out the contrast between the positive and negative peaks, but the use of grey-scale plots, as described below, is an even better solution.

Dot density plotting tends to be a slow process with ordinary pen plotters. This has been overcome by Scollar and Becker in Germany by the use of microfilm plotters, which produce extremely high quality results, but are very expensive.

Fig. 106. Dot density plot of the Maiden Castle survey, made at large scale on a line printer (compare Fig. 107).

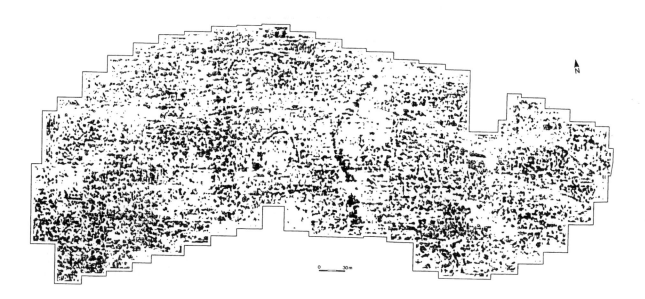

Grey-scale plots

Alongside dot density, computer line printers have been adapted to the grey-scale display of surveys, with the advantage of high speed. They produce, in effect, dot density plots but it is difficult to avoid an assortment of patterning effects at different levels, which can actually obscure the gradations. Overall, there is a certain coarseness in such plots, but some very useful results have been produced in this way (Fig. 106).

Nothing short of a revolution has occurred with the recent appearance of laser printers, and others of comparable refinement (Fig. 107). Now it is possible to create a genuine half-tone picture by means of dots of different sizes, exactly as a photograph is reproduced in a book or newspaper – with the added bonus of publication-quality lettering. Rather than using dots of one size in randomized positions, as with dot density, the half-tone dots are on a regular but very fine grid and are varied in size. This gives a less 'scruffy' effect and cleaner backgrounds which, combined with a continuous grey scale make it possible to show as wide a range of data from low to high as is required, so that positive and negative anomalies can be displayed clearly in a single plot. The type of plot, discussed above, in which normal positive anomalies due to pits and ditches are shown dark, alongside the weak negative anomalies of building foundations in white, is normally more effective than with dot density. Emphasis is added to weak anomalies by the reverse anomalies that accompany them. The full sensitivity range of modern instruments, especially magnetometers, can be fully exploited to reveal subtleties never before seen in archaeological prospecting.

A remarkable aspect of the grey scale is its ability to display fine texture reflecting human activity in relatively blank areas. Grey-scale plans can be looked at as though one were flying over the landscape itself, stripped of its soil. Features are seen in their true positions and, viewed obliquely, different phenomena, especially linear ones, can be revealed as the vantage point is varied, just as with crop marks.

Let us look at some examples in the Maiden Castle plot (Fig. 107). Viewed from the east or west, the two ditches of the eastern end of the long mound stand out clearly, whereas from above, or any other direction, they are invisible.

Although the dominant impression is of an unordered mass of pits, relatively clear strips representing streets can be seen running through them, especially fanning out from the east entrance. The long mound itself is not much encroached upon by pits, and seems also to have served as a road. Other lineations, especially on the northwest, seem due to relatively recent cultivation ridging. The central rounded enclosure is divided by a quite straight line between smooth and rather 'chaotic terrain' (to borrow a phrase from planetary science) suggesting some kind of human usage. Within the latter, there is a hint of a circular trench at the north end, perhaps belonging to a building. The strong north-south feature east of the enclosure is the line of the original bank and ditch defences of the hillfort before it was enlarged, and is visible on the air photograph. A square feature attached to its west side is a recent dewpond. The dot density plot (Fig. 106) shows only the more substantial anomalies: any attempt to bring out more detail by this method would have resulted in confusion.

One must always return to the mundane problem of reproducibility. With a good electrostatic copier, laser half-tone plots are copiable to an adequate standard for normal reports. For quality printing, they reproduce well, as can be seen here, serving as camera-ready copy with the screen necessary for photographic reproduction already applied.

Colour and pseudo-relief

The use of colour in place of density levels in survey plans certainly produces attractive effects, and can be useful for emphasizing particular levels of reading in a plot. However, critics point out that colour cannot be used to show continuous gradations and, in fact, one is actually restricted by the number of colours available to rather coarse steps of contour plotting. Another complaint is that colour is rather flat in appearance compared with modern grey level density plots: it is not adapted to progressive change. However, modestly priced printers are now available which can display refined spectral ranges and promise increasingly subtle assistance in interpretation.

An area in which colour does seem to be clearly effective is in the representation of radar profiles.

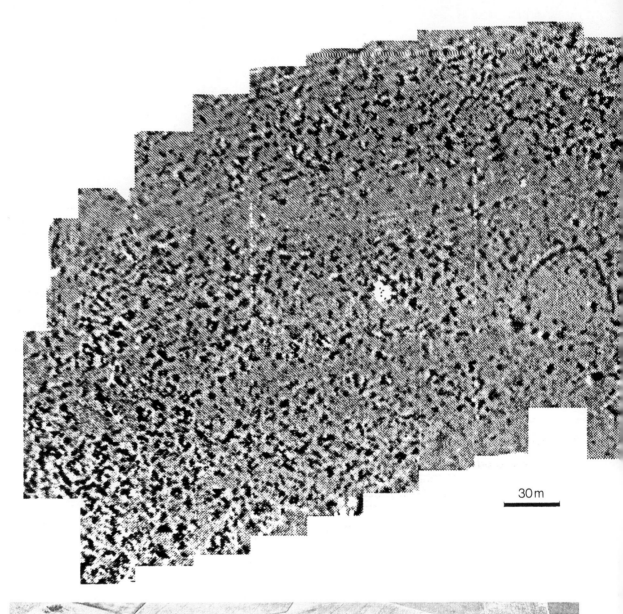

30 m

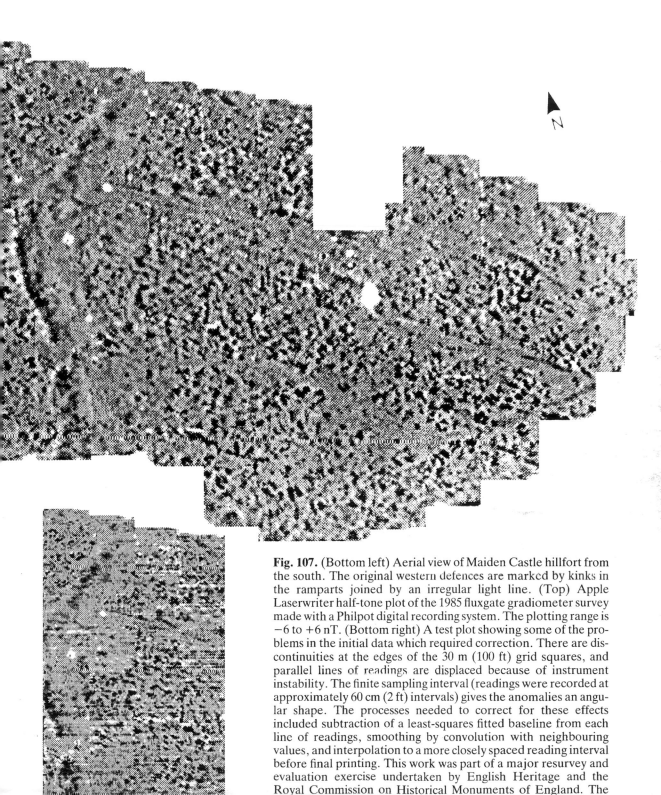

Fig. 107. (Bottom left) Aerial view of Maiden Castle hillfort from the south. The original western defences are marked by kinks in the ramparts joined by an irregular light line. (Top) Apple Laserwriter half-tone plot of the 1985 fluxgate gradiometer survey made with a Philpot digital recording system. The plotting range is −6 to +6 nT. (Bottom right) A test plot showing some of the problems in the initial data which required correction. There are discontinuities at the edges of the 30 m (100 ft) grid squares, and parallel lines of readings are displaced because of instrument instability. The finite sampling interval (readings were recorded at approximately 60 cm (2 ft) intervals) gives the anomalies an angular shape. The processes needed to correct for these effects included subtraction of a least-squares fitted baseline from each line of readings, smoothing by convolution with neighbouring values, and interpolation to a more closely spaced reading interval before final printing. This work was part of a major resurvey and evaluation exercise undertaken by English Heritage and the Royal Commission on Historical Monuments of England. The magnetometer survey covered 13.2 ha (32.6 acres) within the hillfort, involving 132 km (82 miles) of traverses, mostly measured in sub-zero winter conditions. A large area outside the fort was also surveyed. For further discussion see text.

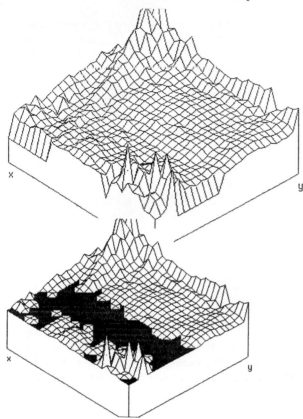

Fig. 108. Farnham Castle perspective plots.

In recent sophisticated work, colours are related to the density and hardness of the materials encountered, with great enhancement of clarity – here one is looking for contrast rather than gradation.

Various pseudo-perspective effects can be produced using available software, or modifications of it. These are sometimes called 'wire frames'. The effective plot of a resistivity survey at Farnham Castle (Fig. 108) was produced by David and Andrew Graham on an Apple Macintosh computer system, with experimental delineation of the moat by (appropriately) 'flooding' the lower reading levels with solid black. An attraction of this type of display is that it can be rotated to find the most telling view. Strata Software of Bradford have specialized in this type of presentation especially for archaeology, and have also combined it with colour.

Filtering and smoothing

Methods that give 'total' measurements, such as resistivity and magnetic susceptibility, have the disadvantage that all readings are superimposed on an arbitrary background level, which may vary from place to place on the site because of geological or soil depth changes for instance, making the clear plotting of a survey difficult to achieve without resorting to mathematical adjustment of the data. This is less important in differential magnetometer or gradiometer surveys, except perhaps for the removal of iron spikes, because these instruments give zero reading in uniform background field; thus they automatically produce level baselines and are, in effect, self-filtering. Fig. 47 (bottom) exemplifies this.

Fig. 104 shows a number of early dot density plots hand-produced by myself and colleagues. All are of ploughed-out square Iron Age barrows on the Yorkshire Wolds. The resistivity plot of Burton Fleming led to the choice of this site for the instrument comparison experiment described in Chapter 1. Alongside it is a plot made at much the same time of the nearby site of Cottam. Here a more sophisticated effect was attempted by the use of a single transfer tint, overlaid upon itself with minor displacements to produce a suitably increasing density. The third plot is a much more rapidly produced, coarse dot density plot of proton magnetometer measurements at Cottam, giving sharper definition than the resistivity. The resistivity plots, which lack the self-filtering attribute, exemplify problems inherent in the use of raw data.

The ditches of the barrows gave low readings with the square array resistivity system, and the dot density has therefore been chosen to increase with decreasing readings. At Burton Fleming, the cut-off level was chosen to give the best definition of the barrow in the south-east quadrant, where the survey was begun. The north-south streaks south-west of this barrow are due to the slight reduction of surface resistivity by intermittent rain during the survey. The survey was resumed in the south-west quadrant after an interval, during which the rain has lowered the general level of resistivity quite uniformly. There was a slight reduction of the rain effect in the two northern quadrants due to some subsequent evapotranspiration, but there is a particularly dark strip along the top of the north-east quadrant which was due, not to rain, but to the extra deep soil of a ploughing headland along the field boundary. The overall effect is that heavy dot density has largely obscured barrows in the upper part of the survey area. The headland phenomenon can also be seen at the bottom of the Cottam survey, where the field boundary ran.

An interesting interpretative detail revealed especially by the Burton Fleming survey is the poor definition of the graves. This must be because they were backfilled with the predominantly chalk gravel natural into which they were cut, with little change of texture or density in the process, while the ditches will have silted slowly and acquired a considerable organic content with good water retention. At Cottam, the graves are artificially emphasized by the effects of earlier excavations.

Step changes between blocks of survey data, like those caused by the rain at Burton Fleming, may be reduced by averaging readings along their corresponding edges, and then adjusting values

in one of the squares by adding or subtracting the difference. The essentially level baselines of magnetic surveys may be affected by instrument drift, which can be checked by returning to a fixed reference point and applying any compensation needed by progressive adjustment of the data; or the baselines can be equalized by the statistical method of least squares fitting.

The background level of surveys can be equalized more subtly by spatial filtering, which may also be used to clarify selectively the representation of archaeological features. Filtering is demonstrated schematically in Fig. 109, which shows a vertical slice through a survey. High readings caused by two wall foundations are visible, but they are superimposed on a steady background increase from left to right caused by deepening topsoil. Choosing a background level as indicated by the broken line will show up the left-hand wall, but the right-hand one will be lost in the high background density. A higher background level suitable for the right-hand wall will not show the left-hand wall at all. If, however, a simple filter is made by subtracting the mean of readings on either side from a central reading and

Fig. 109. The concept of spatial filtering.

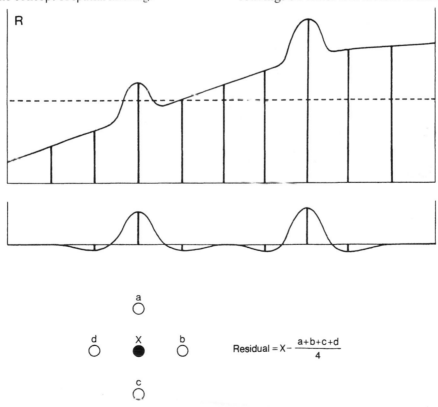

$$\text{Residual} = X - \frac{a+b+c+d}{4}$$

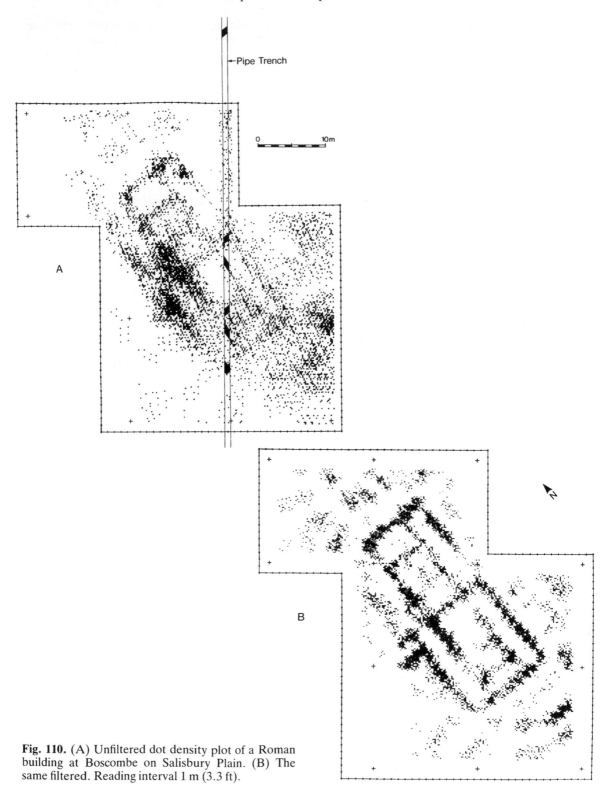

Fig. 110. (A) Unfiltered dot density plot of a Roman building at Boscombe on Salisbury Plain. (B) The same filtered. Reading interval 1 m (3.3 ft).

plotting the so-called residuals that result, the overall slope is largely eliminated and the background levelled, leaving only the residuals representing the walls as significant anomalies. This, in effect, sets up a separate, localized, base level for each reading by comparing it with its neighbours.

The filtered plot shows minor negative peaks on either side of the main ones. These are filter artefacts, which can cause significant distortion in some instances, but in others can be helpful in emphasizing small anomalies, in the same manner as the negative anomalies associated with magnetic peaks. In practice, filters nearly always make use of a ring of readings around the central one, so that they are equally effective whatever the orientation of the buried remains. The art of filter choice, as with any kind of filter, is to select a diameter greater than the width of the features required to pass through it, and to reject anything bigger. The four-element ring filter shown, with a one-reading radius, is the simplest. If the central reading is included in the mean, a slight smoothing effect is obtained, which can be useful with 'noisy' surveys. Another type of filter, called a box filter, uses the mean of a block of readings around the central one. Because it is not tailored to the size of features sought (except that it must be considerably wider) it has a greater smoothing effect, especially on larger features. Straightforward smoothing is obtained by simply averaging groups of readings, or a more sophisticated control can be obtained by giving greater weighting to readings near the centre of the group (see, for example, Fig. 107). Quite modest smoothing can greatly improve the coherence of 'noisy' data.

Fig. 110A shows an early computer-produced dot density plot of a Roman building surveyed with the 0.76 m (2.5 ft) square array resistivity system on a 1 m (3.3 ft) grid, and is a good example of how spatial filtering is applied in a simple practical case. Here the high readings are significant. The existence of the building was first revealed by a series of flint wall foundations on chalk, cut through by a pipe trench. The raw data plot of the survey seemed to show the shadowy plan of an uncomplicated building, consisting of a series of rooms with a corridor along the east side (oddly interrupted by a cross-wall), and a verandah-like structure on the west, downhill, side. As the objective was to clarify the walls, a narrow filter was necessary and the first tried was the simple, four-element, radius 1 filter described above. The angle of the building relative to the survey was particularly suitable for this filter (Fig. 111A), but the residuals were rather small relative to the 'noise' level of the survey, presumably because the outer points were too close to the walls and their associated rubble. An eight-element, diameter 2, filter was more effective (Fig. 112) and was used to produce the final plot (Fig. 110B). The more values used, the less is the chance of the mean being affected too much by one rogue reading, or by those that lie over the feature itself and are thus of similar value to the central reading.

The filtering revealed some remarkable detail, and even made possible some suggestions about the development of the building that were largely confirmed by excavation. Half the 'verandah' was rejected by the filter because it was too wide and the building emerged as consisting of two blocks, one wider than the other and with a porch-like extension on the west side.

The final interpretation was that the building had been of two-phase construction, and that the end wall had been demolished to below floor level when the corridor was extended. It was assumed that the broader southern end had been built first, and the northern part added as a wing. A small selective excavation confirmed the two-phase construction and the deductions about the corridor wall, but showed that the northern part of the house was actually the earlier – demonstrating the danger of subjectivity taking over where the legitimate limits of geophysical survey interpretation are reached.

The excavation also revealed that the part of the 'verandah' that had been filtered out was a mass of rubble used to consolidate a large underlying Iron Age ditch when the building was erected. Furthermore, the filtered plot clearly showed walls that were difficult to distinguish from rubble in the excavation. An intriguing detail was provided by the south-west corner of the building. The sections on either side of the pipe trench seemed to show that there was a wall here with an orientation at variance from the others. The survey, however, revealed that the trench had chanced to catch the corner of the building, so that the two sections, although only a metre apart, were actually of different walls.

Filtering is an obvious task for the computer. However, small surveys like the one described

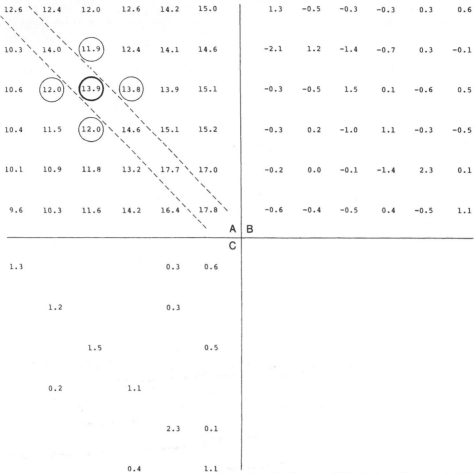

Fig. 111. Filtering applied to measurements of the type obtained from the Boscombe Roman building. (A) Raw data with the filter superimposed. Although simple, this filter is effective for a wall at the angle shown by the broken lines because the four values contributing to the mean are clear of the wall when the central value overlies it. (B) Residuals after filtering. Values at the edges are least reliable, because there are only figures available for part of the filter. (C) Positive residuals only, showing high values over the wall.

above can be tackled with a calculator and hand plotting – and great patience. It is essential to plot out the original numbers with the aid of squared paper, so that a filter template can be laid over them to pick up the necessary figures, as in Fig. 111. The template can be made of paper or card with appropriately spaced holes, or better from a piece of transparent plastic marked with circles. One great advantage of a computer is that it can quickly test the effectiveness of various filters and

density levels. This is hardly practicable for the manual plotter, who must therefore strive to get things right first time.

Band-pass filters In the processing of many sets of survey results, it is necessary not only to filter out the broader variations likely to be due to geology but also to reject small, sharp, often single-reading anomalies that may be due to misreading or poor contact with resistivity, or iron spikes in magnetic surveys. This can be done by subjecting the data to band-pass filtering, which really means processing it with two filters, and only accepting the band of values that passes between them. A narrow filter is first used to remove the spikes, by rejecting the residuals it produces; then a broad filter is used to level out the background (Fig. 113). A band-pass filter was used for the large resistivity survey in the area of the Grime's Graves Neolithic flint mines in Norfolk to remove the broad background effects and indi-

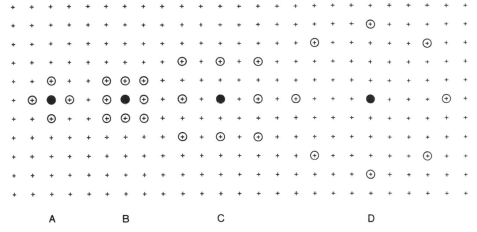

A   B   C   D

Fig. 112. Various ring filters for features of increasing size. The radius 1 filter, A, was effective for the Boscombe building, but C, radius 2, was better.

Fig. 114. Alister Bartlett with the dual spacing (0.5 m and 1 m) twin electrode frame fabricated from electric fence uprights.

vidual false readings; the latter occurred quite often because the dry sandy soil gave very high contact resistances. The band-pass filter was set to pass quite large features running downhill, thought to be linear 'open-cast' flint mines. In retrospect, these are more likely to have been shallow periglacial channels, although they did contain evidence of occupation.

Dual spacing resistivity surveys Scollar long ago pointed out that if a resistivity survey were repeated at two different electrode spacings, both would be similarly affected by the normal resistivity of the ground, but the narrower would

Fig. 113. Band-pass filtering.

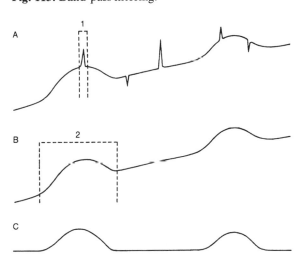

respond more strongly to relatively superficial archaeological remains. One would also expect the shallower reading to be the more responsive because it is looking at a smaller volume of ground and is therefore the less smoothed. Thus, in the absence of archaeology, the ratio or difference between the two resistivity values at each point should be a fairly constant small value compensating for the broader background variations. The effect produced is of vertical filtering, or a resistivity gradiometer.

This elegant concept does not appear to have been followed up until Bartlett and the writer initiated a small experiment at the Prebendal, Thame. An ideal arrangement is a double form of the twin electrode frame with spacings of 0.5 m and 1 m (1.6 ft and 3.6 ft), as shown in Fig. 114. The readings of the twin array are proportional to aR (p.46), therefore over uniform ground the 1 m (3.3 ft) reading is exactly half that for 0.5 m (1.6 ft), and the 0.5 m (1.6 ft) reading minus twice the 1 m (3.3 ft) reading will be zero. Any deviation from this relationship will indicate an anomalous response from one of the two spacings, most likely the narrower.

The site is rather complex and perhaps not ideal for a first experiment, but an example from one series of readings does support the promise of this approach (Fig. 115). A background trend is removed and the baseline is levelled. The fact that the baseline is well below zero shows that the resistivity increases with depth, which is in keeping with the likelihood that the subsoil is well-drained gravel.

Symbol plots

There are some surveys which produce results so imprecisely defined, or 'noisy', that none of the above techniques will give a comprehensible or visually acceptable picture. This is also often the case when data points are widely spaced and possibly therefore too tenuously related for a treatment which tries to combine them to be acceptable. Such readings are best represented by symbols proportional in size to the readings. The results from the Tadworth enclosure were such a case (Fig. 85), where neither contouring nor dot density produced a coherent plan from the noisy and widely-spaced data. However, symbol plotting proved highly effective, the eye being able to recognize the crude clustering of the sym-

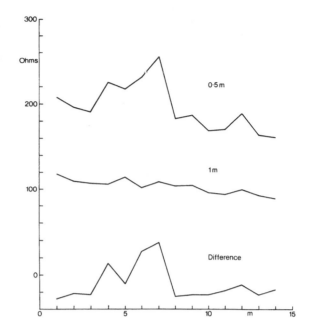

Fig. 115. Dual spacing resistivity traverse made with the twin electrode configuration shown in Fig. 114.

bols without being encouraged to be critical of their variations in size, or even of the occasional completely disparate value. This can be compared with the Dainton surveys (Fig. 88), in which data at the same ground spacing are represented with shaded squares; these are also effective and quite easy to produce, but are vulnerable to filling in when reduced for printing.

The magnetic susceptibility survey of Coneybury (Fig. 86) was, by contrast, closely measured but symbols were again effective for more complicated reasons. Although an acceptable dot density plot was produced by careful filtering and smoothing, it was found that the use of white and black symbols made it possible to draw attention to significant low readings without detracting from the visual impact of the more important simple central cluster of high readings. The enlargement of the symbols towards the more extreme values also avoided the distracting effect of the central clutter of medium-value readings obtained with dot density. Furthermore, the rather restricted range of the readings made it necessary to amplify them for plotting, emphasizing their rather high 'noise' level, so that, as in the coarse survey case, symbols were appropriate for achieving the most coherent effect.

The methods introduced in this chapter belong to the science and art of image processing, which continues to grow in sophistication and subtlety, especially in the areas of space imagery and large-scale geophysics. Archaeology has derived much benefit from developments stimulated by these modern applications, and contributes a little in return.

Chapter 8
Survey logistics

This brief chapter introduces the basic methods of site preparation and survey procedure that are common to most prospecting techniques. Procedures specific to individual techniques are discussed in the relevant chapters. Rather than direct Imperial equivalents to metric measurements, suitable practical alternatives are given where appropriate. It would have been attractive to recommend a subdivision of 100 ft as the standard Imperial module, but the metric field system is tied to the 1-m module for functional reasons; therefore 3 ft and its multiples and submultiples are the most practical choice.

Grids

Most types of survey require a grid to be marked out, related to permanent features on the ground and the National Grid so that it can be readily re-established for excavation or any other subsequent use of the survey information. Grids are the essential foundation of area surveys, and can also be used as the basis for non-intensive magnetic susceptibility and phosphate sampling and resistivity or conductivity search surveys.

The standard size of grid square used by the Ancient Monuments Laboratory for many years is 30 m, really a translation of 100 ft (30.48 m) into its closest metric equivalent. Another frequently used standard mesh is 20 m (60 ft). A disadvantage of 30 m is that it can be difficult to fit into areas of awkward, confined shape, but we have generally found this outweighed by the economy of the laying-out process and the moving on of equipment.

All surveys should be related to the National Grid, and it is good practice to use it as the basis for the actual survey grid, if convenient. It has the advantage, in survey reports, of enabling one to specify the locations of detected features by their National Grid references, so that they are fixed in space by a generally recognized co-ordinate system which facilitates direct comparison with other types of survey and sites and monuments records. As the Ordnance Survey continues to develop digitized maps, direct plotting on to these will be possible. It is also independent of the disappearance of any landmarks or pegs that the grid must otherwise be based upon.

In many fields, however, especially smaller ones and those with boundaries at even a quite modest angle to north, the direct use of the National Grid can be awkward and wasteful of space (Fig. 116). With surveys using separate readings, the odd corners can be filled in, but with automatic recording it can be difficult to cover all corners. It is often more convenient to use a grid fitted to the field (Fig. 117), and relate it to the National Grid by giving the Grid reference of two points to establish its position unambiguously. In Fig. 116, the references for the junction of the two baselines and one end of the main baseline are given. The 14-figure references, giving the locations to the nearest centimetre (0.4 in.), were only possible because this plan was based on a 1:1250 map. Even so, it might be regarded as overkill, not to say optimistic; but it is good practice to aim for over- rather than under-precision. Modern electronic survey equipment can give even better resolution within the limits of a particular survey, but the usefulness of this precision in linking the survey to the outer world is limited by the accuracy with which the National Grid can be established on the ground. If the 1:2500 map is used as the base, a 12-figure reference is amply sufficient. The survey should also be measured in to permanent ground reference points – gateposts, fence corner posts or buildings, for instance – which may be more convenient to use on the ground than the National Grid, as well as providing a check on the Grid references. Permanent pegs are also important for marking the positions of baselines; these should be well hammered in, preferably in fairly obscure positions under

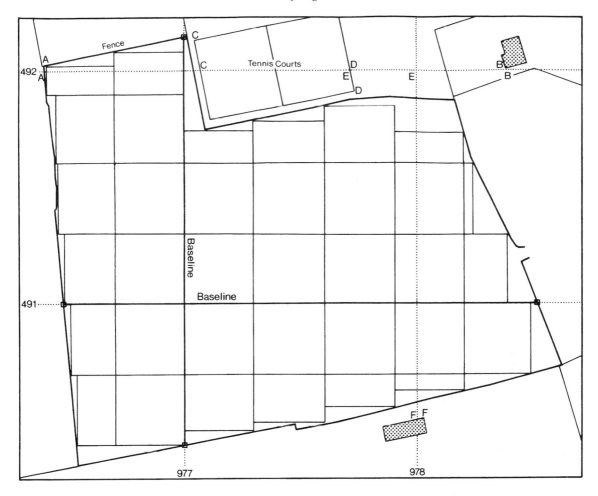

Fig. 116. Survey grid of 30 m (100 ft) squares based on the National Grid.

hedges, etc., where they will not attract the attention of those with a compulsion to remove things. For long-term durability, they should be concreted in. The importance of good locational references cannot be overemphasized: archaeologists have been known to find a geophysical survey useless because it was inadequately or unclearly related to the rest of the world.

If the National Grid is to be used as the direct basis of the survey grid, it has first to be established on the ground. The only practical way of doing this for everyday work – although it may not look very rigorous – is to use an Ordnance Survey map or plan, preferably of 1:2500 scale or greater, to ascertain where the 100-m grid lines cross, or come within accurately measurable dis-

tance of, clear-cut permanent features such as walls and houses. If nothing else is available, less precise measurements may have to be obtained from hedges. It may be necessary to go some distance outside the survey area to find suitable points. Fig. 116 provides an example. Grid line 492 can be established by measuring the distance BB from the corner of the house, and AA from the end of the fence. Although this latter measurement is along a hedge, the critical thing is the position of the fence, which is a precise feature. The accuracy of the line thus established can be checked by measuring CC and DD along the fences of the tennis courts. If ranging poles are put at each of the measured points, they should all line up. The position of the line 978 can be determined by measuring EE from the tennis court, and establishing a right-angled line from

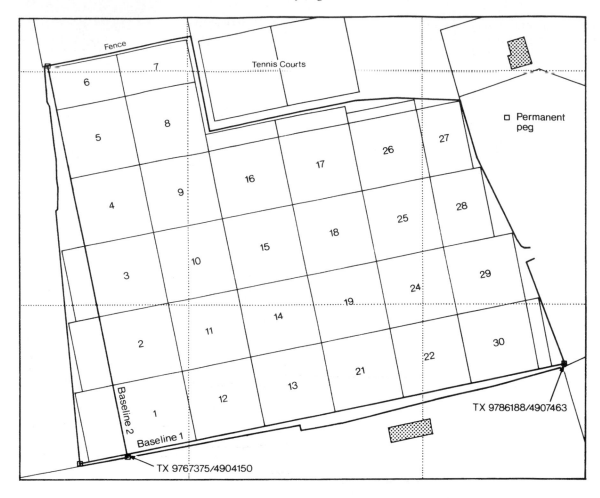

Fig. 117. Survey grid of 30 m (100 ft) squares fitted to the field.

this point, which can be checked by measuring FF along the wall of the building to the south. The rest of the grid can be built up by measuring from the two lines thus established.

The grid fitted to the field leaves a number of partial squares; the grid based on the National Grid leaves rather more, as well as some triangular areas. Filling in such areas can be time consuming and difficult, especially with survey systems involving tripods and strings, and in practice some at least are often left out. Their absence rarely has much effect on the usefulness of the survey, as the area covered is usually a quite comprehensive sample of the field, giving strong hints of the distribution of remains even beyond its

limits. The strips close to the tennis courts in Fig. 117 are left out on the assumption that the survey is a magnetic one and their fencing would cause interference.

Laying out the grid It is very useful to have a grid of appropriate scale drawn on a transparent sheet such as a piece of tracing paper, which in Britain can be laid over the 1:10,000, 6-inch, 1:2500 or 1:1250 OS plan of the area to be covered. As well as being a great help in planning an actual survey, this enables one to calculate the work and time that will be involved. Having used this for pre-planning, one can often begin laying out the survey grid on site after only a brief reconnaissance to ascertain whether unmapped hazards such as temporary fencing or abandoned farm machines are present.

In the preliminary planning, and on site, a baseline must be established as close as possible to the longest edge of the survey area, or the longest unencumbered National Grid line (Figs. 116, 117), making allowance, if necessary, for the overlap of any distance transducer tripods if the survey is to run along the baseline. This baseline should be marked with ranging poles. If a theodolite is being used, a secondary baseline should be established at right angles to the first, again where it is possible to achieve the longest straight run compatible with the largest number of complete squares.

Once a single accurate right angle has been established between the two baselines, the grid can be built up entirely by measuring offsets from these. The right angle can be obtained by means of a 3, 4, 5 triangle, using the longest dimensions the available tape or tapes allow, but preferably a theodolite. Alternatively, a very effective method of laying out grids of moderate size without a theodolite is to use an optical square – an inexpensive little hand-held instrument, normally prismatic in principle and requiring no setting-up process. It enables one to establish right angles and to interpolate along straight lines, sighting on vertical markers such as ranging poles. An accuracy within ± 3 cm (1 in.) at 30 m (90 ft) can be readily achieved, which is normally well within the tolerance required of a geophysical survey. Using an optical square, the best way to lay out the grid is with a series of right-angled offsets from the baseline at the chosen grid intervals, and then to measure similar lengths along the offsets. This approach ensures that any errors are not perpetuated or accumulated in the grid, and is also ideal for surveys, such as search traverses, based on linear offsets from the baseline.

If the grid is to be long-lasting, and there is no danger to livestock, it is best marked out with substantial wooden pegs, well hammered in; these should be 5 cm (2 in.) square and at least 30 cm (1 ft) long. If it is to be taken up immediately after the survey, 3–4 ft (about 1 m) garden bamboos, or even smaller, can be used. To improve their visibility, these should be coloured at the top with fluorescent paint or self-adhesive tape; or very good flags can be made with ordinary brightly coloured tape obtainable from upholstery shops.

Survey procedures

The procedures for a normal 1 m (3 ft) detailed survey are shown in Fig. 118, which is based on a diagram in the Geoscan manual. The method applies to squares of any size, but a 10 m (30 ft) square is shown for clarity. The area is divided into notional 1 m (3 ft) squares, each containing a reading. A and B are tapes or, preferably, strings made from plastic-coated clothes line with measuring points marked with waterproof adhesive tape in alternating bright colours (green should be avoided!), and looped at the ends for attachment to pegs, bamboos or land arrows. Clothes line thus prepared has advantages of clarity and cheapness compared with measuring tapes.

For surveys with separate readings, a third string C, marked at 1 m (3 ft) intervals, is stretched across. The survey follows the arrowed dashed line, each reading being taken beside a marker point, at an estimated 0.5 m (1.5 ft) from the string. One string position serves for two rows of readings, and it is then moved on 2 m (6 ft) to position D for the next two rows. This string also makes a suitable guide for the Geoscan ST1 fluxgate gradiometer recording system, the objective of the instrument carrier being to reach a marker as each timed bleep occurs. Normally, this system will be set to record readings at a closer interval than the 1 m (3 ft) shown. With systems that use a distance transducer, the marked string will be replaced by a raised string supported by two tripods, which are shown dotted.

The double-scale sketch below shows part of string C. The lower figure is surveying with a fluxgate gradiometer, using the width of his body as a guide to keep the instrument 0.5 m (1.5 ft) from the string; his left hand is well placed for the distance transducer on a tripod system. The upper figure is using a twin electrode frame in the single-handed 'walking stick' manner. The attitude of the body when using this is not so suitable as a 0.5 m (1.5 ft) gauge, so he walks on the other side of the instrument and judges the distance from the string. Note, too, that as he is right-handed the survey must be conducted in the opposite direction to that shown above (see also Fig. 35).

An alternative to moving string C after every two traverses is to have two strings, or even a complete set to cover the square, laid out beforehand. This is particularly worthwhile for oneman surveying. It is less important when surveying with a fluxgate gradiometer with a team of two: as this is a one-handed instrument, it is easy to move on string C, the instrument operator picking up one end and his assistant the other. It is, of course, essential to operate in this way with a tripod-supported distance transducer.

Bold and experienced operators using timed walking systems such as the ST1 sometimes make great use of interpolation, replacing string C with bamboos along A and B up to 5 m (15 ft) apart, between which they are able to interpolate at 1 m (3 ft) intervals, and rely on well-practised pacing to get from A to B in the correct time. It is helpful to mark the bamboos similarly on either side of the survey strip, for instance with 1, 2, 3 and 4 coloured tape bands, then 3, 2 and 1, sufficient for a 30 m (90 ft) square; or different colours can be used. It is even possible to position the bamboos by pacing, so that no measurements are made on site after the initial setting out of the grid. Many will find this minimal marking out approach is best used with the smaller 20 m (60 ft) grid size, to reduce errors (see Addendum, p.171).

It is possible to carry out ground-contact surveys with strings 3 m or 4 m (9 ft or 12 ft) apart, but a major problem with this degree of interpolation is the difficulty of viewing the string markers obliquely, especially in long grass or other vegetation. With the two-traverse method first described here, visibility of the string is good because one is looking down at it.

Other survey grid systems have been proposed. For instance, Atkinson pointed out that a square-based grid does not give even coverage of the ground. He advocated a procedure with alternate lines of readings staggered, but with a spacing between the lines of a$\sqrt{3}/2$, where a is the electrode spacing along the line, that is about 0.87 m for 1 m spacing (2.6 ft for 3 ft spacing). In this way, all inter-electrode spacings become equal. This has not been generally adopted because the small gain in evenness of coverage is outweighed by the difficulties of using such measurements with standard grids.

Extra-detailed surveys If a survey with separate readings at 0.5 m (1.5 ft) interval is required, a convenient approach is to measure at 0.5 m (1.5 ft) intervals along the cross-string C, but to keep the interval along A and B at 1 m (3 ft) and then repeat the survey staggered upward by 0.5 m (1.5 ft), combining the results at the processing stage.

Halving the spacing in this way increases the number of readings in a 30 m square to 3600 – four times as many as with 1 m (3 ft) spacing but not necessarily four times as productive, and often prohibitive in time and effort.

A good compromise, which only doubles the amount of work of a 1 m (3 ft) survey, is shown at the bottom of Fig. 118. The reading interval along C is kept at 1 m (3 ft), but the traverses are 0.5 m (1.5 ft) apart and staggered, which in effect produces a square grid orientated at 45 degrees, with a side of about 0.7 m (1/$\sqrt{2}$ m), or 2.1 ft. This is not much larger than 0.5 m (1.5 ft), yet

Fig. 118. (Top) Standard survey procedure. (Middle) Operating a twin electrode resistivity frame (right) and a fluxgate gradiometer with the grid. (Bottom) Procedure for the approximate 0.7 m (2.3 ft) grid arrangement.

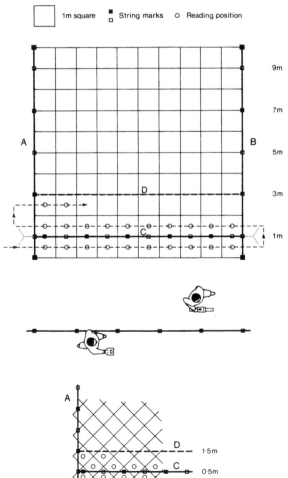

only twice the density of readings of a 1 m (3 ft) grid. With some adjustment and interpolation, this can be done with the same strings as the 1 m (3 ft) survey, readings being taken 0.25 m (10 in.) from string C, alternately opposite the markers and between them. The first reading must also be 0.25 m (10 in.) from string A to ensure the correct fit between adjacent squares. Alternatively, two suitably staggered 1 m (3 ft) surveys can be super-imposed as above. This interval proved very effective for the magnetic susceptibility survey of Coneybury henge (Fig. 86).

This scheme may prove difficult to use because the effectively diagonal grid it produces is not con-veniently suited to most software packages. However, such a grid can be replaced by the normal rectilinear form while retaining comparable reading intervals, although it will be necessary to prepare specially marked strings for this.

The fundamental need for a study of reading interval optimisation has been surprisingly neglected, probably because of the pervasive need to cover the ground economically. The standard ground reading interval of 1m for resistivity has long been suspected of not being ideally close, while gradiometer coverage is uneven – as intense as necessary along traverses, but with traverses nor-mally 1m (say 3 ft) apart. The problem is now being addressed by A. Schmidt at Bradford University, who has demonstrated that, for example, forecourt features at the Hazleton long barrow in the Cotswolds, mostly in the form of narrow ditches, were detectable at 0.5m (1.5 ft) resistivity spacing (0.5m twin electrode), but scarcely at 1m. A prob-ably Saxon building in the medieval village of Swavesey, on the edge of the Cambridgeshire fenland, was clearly defined at 0.5m (1.6 ft) spacing: 'If the survey had been done at the usual reading interval of 1m, it is likely that these tenuous but important features would have been missed.' (Maekawa *et al*, 1995). A spacing of 0.25m (0.8 ft) does not appear greatly to improve the reso-lution. These findings are in keeping with Fig. 62 of this book and its accompanying argument.

A solution for magnetometry is to use multiple sensors. The Institut für Geophysik of the Christian Albrechts Universität, Kiel (who provide a comprehensive survey service), conduct intensive surveys by means of five gradiometers mounted at 0.2m (0.66 ft) spacing across the front of a wooden cart. Such an arrangement also allows the option of having the sensors more widely separated for faster ground coverage. For resistivity, the solution lies in multiple probe frames such as the Geoscan PA5, or possibly multiple-wheeled systems (cf. Fig. 47).

Non-intensive surveys. (Fig. 119) In the context of this book, these are almost always magnetic susceptibility surveys, although the principles can be applied to phosphate sampling or even field-walking. The concept of dividing a large grid square into a number of cells is the same as that shown in Fig. 118 (top), now the standard approach for surveys of all types and sizes. Marking out has been reduced to a minimum.

Typically, areas are first marked out with a 100m (say 300 ft) grid. This is conveniently done by means of right-angled offsets from a roughly central baseline, aligned with the National Grid or site grid when suitable. Initially the grid can be marked with pegs or fairly inconspicuous short flagged bamboo canes, especially in areas with public access; these can be made more visible square by square with ranging poles for the actual survey. This is done simply by pacing appropriate distances along one side of the square, and then setting off across the square, estimating the direc-tion in relation to the further ranging poles and sampling at paced intervals, normally of 20m (60 ft) or 10m (30 ft). In this way, a whole square is surveyed or sampled without any further marking out, although it is useful to have a flag or pole which can be left at the beginning of each outward traverse to help in judging the direction of the return tra-verse; or preferably to use the pole-guidance scheme shown in Fig. 68.

This procedure seems at first sight to be a very imprecise approach to measurement, but it is only essential for a reading to lie somewhere within a cell 20m (60 ft) or 10m (30 ft) square, and few people find difficulty in achieving this, especially if they are able to cultivate a 1 m (3 ft) pace. To maintain accuracy at smaller reading intervals, say 5m (15 ft) or less, the 100m (300 ft) squares can be subdivided into, or replaced by, 50m (150 ft) or even 25m (75 ft) squares; or one can work entirely in 30m (say 100 ft) or 20m (60 ft) modules, espe-cially if the survey is to be followed by a magnetometer survey over the same area.

For narrow linear surveys, a central or lateral baseline is necessary, along which offsets can be positioned. These will normally be short enough not to need aligning by measurement.

Only one person is essential for surveys of this type, once the baseline and offsets have been laid out.

Gradiometer survey update. When one is relying on the ST1 timed trigger device combined with accurate pacing for precision work in adverse conditions such as desert scrub or plough furrows on heavy soils, it is difficult to maintain accurate pacing for more than 10m (30 ft). However, strings 10m (about 30 ft) apart stretched across the survey squares at right angles to the traverse lines provide effective and economical controls, enabling squares of any size to be accurately surveyed. A 30m (100 ft) square requires two such strings, and a 20m (60 ft) square only one. For even better accuracy of both distance and direction, one can walk along either side of a calibrated string laid on the ground, in the same manner as with the tripod and potentiometer system (pp. 71–6), but this does require a second person.

Schemes based on individual 30m (100 ft) or 20m (60 ft) squares (Figs 116, 117) are now often replaced by continuous survey strips 100m (300 ft) long, which are especially suitable for the linear surveys now required for potential communication corridors, and the alternate strip sampling method. Plastic-coated lines are linked to produce the 100m (300 ft) length, and then laid parallel on the ground at the appropriate 20m (60 ft), 30m (100 ft) or even greater width. The instrument operator walks between markers on the strings, normally using the ST1. The markers are conveniently made from alternate black and white plastic kitchen bin liners. Experienced workers can interpolate 1m (3 ft) traverses between these when they are placed at 5m (15 ft) intervals, but an interval of 4m (12 ft) is rather less demanding on the attention.

Again, such a survey can be undertaken by one person once the essential grid of 100m (300 ft) × selected width has been marked with pegs or canes.

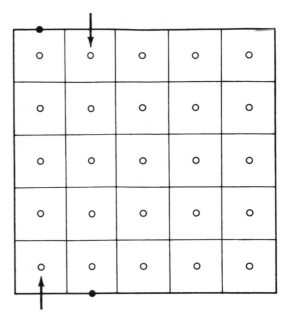

Fig. 119. Scheme for a sampling survey, for example magnetic susceptibility at 20 m intervals over a 100 m (1 ha) square. Dark circles are marker ranging poles which are moved on as the survey progresses, as in Fig. 68. Optimum sampling positions are shown with open circles. Measurements within the square are paced. The layout and results of such a survey are shown in Fig. 124, for which the baseline was the 200 mN grid line.

References

AITKEN, M. J., 1959. *The Listener* 61, 325.

AITKEN, M. J., 1974. *Physics and Archaeology*, 2nd edition. Clarendon Press, Oxford.

AITKEN, M. J. and TITE, M.S., 1962. 'A gradient magnetometer using proton free precession.' *Journal of Scientific Instruments* 39, 625–629.

AITKEN, M. J., WEBSTER, G. and REES, A., 1958. 'Magnetic prospecting.' *Antiquity* 32, 270–271.

AL CHALABI, M. M. and REES, A. I., 1962. 'An experiment on the effect of rainfall on electrical resistivity anomalies in the near surface.' *Bonner Jahrbucher* 162, 266–271.

ALLDRED, J. C., 1964. 'A fluxgate gradiometer for archaeological surveying.' *Archaeometry* 7, 14–19.

ARRHENIUS, O., 1929. 'Die Phosphatfrage.' *Zeitchrift fur Pflanzenernahrung, Dungung und Bodenkunde* 14, 185–197.

ASPINALL, A. and LYNAM, J. T., 1970. 'An induced polarization instrument for the detection of near-surface features.' *Prospezioni Archeologiche* 5, 67–75.

ATKINSON, R. J. C., 1952. 'Méthodes électriques de prospection en archéologie.' In LAMING, A. (ed.), *La découverte du passé*, 59–70. Picard, Paris.

ATKINSON, R. J. C., 1953. *Field Archaeology*. 2nd edn. Methuen, London.

ATKINSON, R. J. C., 1963. 'Resistivity surveying in archaeology.' In Pyddoke, E. (ed), *The Scientist and Archaeology*, 1–30. Phoenix House, London.

BACON, E., 1960. *Digging for History*. A. & C. Black, London.

BAKER, R. R., 1989. *Human Navigation and Magnetoreception*. Manchester University Press, Manchester.

BAILEY, R. N., CAMBRIDGE, E. and BRIGGS, H. D., 1988. *Dowsing and Church Archaeology*. Intercept, Wimborne.

BINTLIFF, J., forthcoming. 'Trace metal accumulations in soils on and around ancient settlements in Greece.' BAR proceedings of the meeting 'Geoprospection in the archaeological landscape', Dorset Institute of Higher Education, 1989.

BOTT, M. H. P., 1971. *The Interior of the Earth*. Edward Arnold, London.

BOUCHER, A. R., 1987. 'Geophysical survey of a Roman fort site using earth resistivity: a comparison of two methods.' B.Sc. dissertation, University of Bradford, unpublished.

BOWEN, H. C., 1975. 'Air photography and the development of the landscape in central parts of southern England.' In WILSON, D. R. (ed.) *Aerial Reconnaissance for Archaeology*. Research Report 12, 103–118. Council for British Archaeology, London.

CARVER, M. O. H. (ed.), 1986. *Bulletin of the Sutton Hoo Research Committee* 4.

CATHERALL, P. D., BARNETT, M. and MCCLEAN, H. (eds), 1984. *The Southern Feeder*. British Gas Corporation, London.

CLARK, A. J., 1957. 'The transistor as the archaeologist's latest tool.' *Illustrated London News* 230, 900–901.

CLARK, A. J., 1980. 'Archaeological detection by resistivity.' Ph.D thesis, University of Southampton, unpublished.

CLARK, A. J., 1983. 'The testimony of the topsoil.' In MAXWELL, G. S. (ed.) *The Impact of Aerial Reconnaissance on Archaeology*. Research Report 9, 128–135. Council for British Archaeology, London.

CLARK, A. J., 1986. 'Archaeological geophysics in Britain.' *Geophysics* 51, 1404–1413.

CLARK, A. J. and HADDON-REECE, D., 1973. 'An automatic recording system using a Plessey fluxgate gradiometer'. *Prospezioni Archeologiche* 7–8, 107–113.

CLARK, A. J., TARLING, D. H. and NOËL, M., 1988. 'Developments in archaeomagnetic dating in Britain.' *Journal of Archaeological Science* 15, 645–667.

COLANI, C., 1966. 'A new type of locating device. I – The instrument.' *Archaeometry* 9, 3–8.

COLANI, C. and AITKEN, M. J., 1966. 'A new type of locating device. II – Field trials.' *Archaeometry* 9, 9–19.

CRADDOCK, P. T., GURNEY, D., PRYOR, F. and HUGHES, M. J., 1985. 'The application of phosphate analysis to the location and interpretation of archaeological sites.' *Archaeological Journal* 142, 361–376.

DARWIN, C., 1883. *Vegetable Mould and Earth-worms*. John Murray, London.

DELETIE, P., LEMOINE, Y., MONTLUCON, J. and LAKSHMAN, J., 1988. 'Discovery of two unknown pyramids at Saqqarah (Egypt) by a multimethod geophysical survey.' Presentation at SEG Annual Meeting, Anaheim.

EIDT, R. C., 1977. 'Detection and examination of anthrosols by phosphate analysis.' *Science* 197, 1327–1333.

ENTWISTLE, R. and RICHARDS, J., 1987. 'The geochemical and geophysical properties of lithic scatters.' In BROWN, A. G. and EDMONDS, M. R. (eds) *Lithic Analysis and Later British Prehistory*. BAR British Series 162, 19–37. British Archaeological Reports, Oxford.

FRIESINGER, H. and DEVAUX, R., 1983. 'Vorbericht über die österreichischen Ausgrabungen auf St Lucia, Westindien.' *Anzeiger der phil.-hist. Klasse der österreichischen Akademie der Wissenschaften* 120, 227–238.

FROHLICH, B. and LANCASTER, W. J., 1986. 'Electromagnetic surveying in current Middle Eastern archaeology.' *Geophysics* 51, 1414–1425.

GINGELL, C., 1982. 'Excavation of an Iron Age enclosure at Groundwell Farm, Blunsdon St Andrew, 1976–7.' *Wiltshire Archaeological Magazine* 76, 33–75.

GINGELL, C. J. and SCHADLA-HALL, R. T., 1980. 'Excavations at Bishops Cannings Down, 1976.' In HINCHLIFFE, J. and SCHADLA-HALL, R. T. (eds) *The Past under the Plough*, 109–113. Department of the Environment, London.

GORMAN, M., 1985. 'Beowulf in 3D – soil sounding radar surveys at Sutton Hoo.' *Seismic Images* 8, 24–29.

GREENE, K., 1983. *Archaeology: An Introduction*. Batsford, London.

HALL, E. T., 1962. 'Some notes on the design and manufacture of detector heads for proton magnetometers.' *Archaeometry* 5, 139–145.

HAMEL, J. V. and JACOBSON, R. B., 1988. 'Archeogeological investigation of a Monongahela River terrace.' MARINOS, P. G. and KOUKIS, G. C. eds *Engineering Geology of Ancient Works, Monuments and Historical Sites* 2, 1163–1171. Balkema, Rotterdam.

HAMPTON, J. N., PALMER, R. and CLARK, A. J., 1977. 'Implications of aerial photography for archaeology.' *Archaeological Journal* 134, 157–193.

HEATHCOTE, C., 1983. 'Applications of magnetic and pulsed induction methods to geophysical prospection at shallow depth.' Ph.D thesis, University of Bradford, unpublished.

HESSE, A., 1966a. *Prospections géophysiques à faible profondeur*. Dunod, Paris.

HESSE, A., 1966b. 'The importance of climatologic observations in archaeological prospecting.' *Prospezioni Archeologiche* 1, 11–13.

HESSE, A., JOLIVET, A. and TADDAGH, A., 1986. 'New prospects in shallow depth electrical surveying for archaeological and pedological applications.' *Geophysics* 51, 585–594.

HINCHLIFFE, J., 1986. 'An early medieval settlement at Cowage Farm, Foxley, near Malmesbury.' *Archaeological Journal* 143, 240–259.

IMAI, T., SAKAYAMA, T. and KANEMORI, T., 1987. 'Use of ground-probing radar and resistivity surveys for archaeological investigations.' *Geophysics* 52, 137–150.

LAMBRICK, G., 1988. *The Rollright Stones: Megaliths, Monuments and Settlement in the Prehistoric Landscape*. Historic Buildings and Monuments Commission, London.

LE BORGNE, E., 1955. 'Susceptibilité magnétique anormale du sol superficiel.' *Annales de Géophysique* 11, 399–419.

LE BORGNE, E., 1960. 'Influence du feu sur les propriétés magnétiques du sol et du granite.' *Annales de Géophysique* 16, 159–195.

LERICI, C. M., CARABELLI, E. and SEGRE, E., 1958. 'Prospezione geofisiche nella zona archeologica di Vulci.' *Quaderni di Geofisica Applicata* 19.

LIMBREY, S., 1975. *Soil Science and Archaeology*. Academic Press, London, New York and San Francisco.

LININGTON, R. E., 1964. 'The use of simplified anomalies in magnetic surveying.' *Archaeometry* 7, 3–13.

LININGTON, R. E., 1967. 'A short geophysical campaign carried out at Bolonia, Cadiz.' *Prospezioni Archeologiche* 2, 49–71.

LININGTON, R. E., 1973. 'A summary of simple theory applicable to magnetic prospecting in archaeology.' *Prospezioni Archeologiche* 7–8, 9–84.

LITTLER, N. S., 1977. 'A comparison of phosphate testing and magnetic viscosity surveying on archaeological sites.' B.Sc. dissertation, University of Surrey, unpublished.

LYNAM, J. T., 1970. 'Techniques of geophysical prospection as applied to near surface structure determination.' Ph.D thesis, University of Bradford, unpublished.

MCHUGH, W. P., SCHABER, G. G., BREED, C. S. and MCCAULEY, J. F., 1989. 'Neolithic adaptation and the Holocene functioning of Tertiary palaeodrainages in southern Egypt and Northern Sudan.' *Antiquity* 63, 320–336.

MULLINS, C. E., 1977. 'Magnetic susceptibility of the soil and its significance in soil science – a review.' *Journal of Soil Science* 28, 223–246.

NOËL, M., forthcoming. 'Electrical resistivity tomography for imaging archaeology.' BAR proceedings of the meeting 'Geoprospection in the archaeological landscape', Dorset Institute of Higher Education, 1989.

OTTOW, J. C. G. and GLATHE, H., 1971. 'Isolation and identification of iron reducing bacteria from gley soils.' *Soil Biology and Biochemistry* 3, 43–55.

OZAWA, K. and MATSUDA, M., 1979. 'Computer assisted techniques for detecting underground remains based on acoustic measurement.' *Archaeometry* 21, 87–100.

PACKARD, M. and VARIAN, R., 1954. 'Free nuclear induction in the Earth's magnetic field.' *Physical Review* 93, 941.

PALMER, L. S., 1960. 'Geoelectrical surveying of archaeological sites.' *Proceedings of the Prehistoric Society*, 26, 64–75.

PERISSET, M. C. and TABBAGH, A., 1981. 'Interpretation of thermal prospection on bare soils.' *Archaeometry* 23, 169–187.

PHILPOT, F. V., 1973. 'An improved fluxgate gradiometer for archaeological surveys.' *Prospezioni Archeologiche* 7–8, 99–105.

POCOCK, J. A. (ed.), 1983. *Geophysical Surveys 1982.* School of Archaeological Sciences, University of Bradford.

POWELL, H. M., BARBER, D. C. and FREESTON, I. L., 1987. 'Impedance imaging using linear electrode arrays.' *Clinical Phys. Physiolog. Meas.* 8, 109–118.

PROUDFOOT, B., 1976. 'The analysis and interpretation of soil phosphorus in archaeological contexts.' In DAVIDSON, D. A. and SHACKLEY, M. L. (eds) *Geoarchaeology* 93–113. Duckworth, London.

RAHTZ, P., HAYFIELD, C. and BATEMAN, J., 1986. *Two Roman Villas at Wharram Le Street.* York University Archaeological Publications, 2.

RALPH, E. K., 1964. 'Comparison of a proton and a rubidium magnetometer for archaeological prospecting.' *Archaeometry* 7, 20–27.

RCAHM, 1982. 'Iona: An Inventory of the Monuments.' *Argyll* 4. Edinburgh: Royal Commission on the Ancient and Historic Monuments of Scotland.

RITCHIE, J. N. G., 1978. 'The Stones of Stenness, Orkney.' *Proceedings of the Society of Antiquaries of Scotland* 107, 1–60.

RUSSELL, E. J., 1957. *The World of the Soil.* Collins, London.

SCHWARZ, G. T., 1967. 'Prospecting without a computer in southern Switzerland.' *Prospezioni Archeologiche* 2, 73–80.

SCOLLAR, I., 1959. 'Einführung in die Widerstandsmessung, eine geophysikalische Methode zur Aufnahme von archäologischen Befunden unter der Erdoberfläche.' *Bonner Jahrbücher* 159, 284–313.

SCOLLAR, I. and KRÜCKEBERG, F., 1966. 'Computer treatment of magnetic measurements from archaeological sites.' *Archaeometry* 9, 61–71.

SIEVEKING, G. DE G., LONGWORTH, I. H., HUGHES, M. J., CLARK, A. J. and MILLETT, A. 'A new survey of Grime's Graves.' *Proceedings of the Prehistoric Society* 39, 182–218.

SORBY, H. C., 1879. 'On the structure and origin of limestone.' *Quarterly Journal of the Geological Society of London* 35, 56.

SOWERBUTTS, W. T. C. and MASON, R. W. I., 1984. 'A microcomputer based system for small-scale geophysical surveys.' *Geophysics* 49, 189–193.

STOVE, G. C. and ADDYMAN, P. V., 1989. 'Ground probing impulse radar: an experiment in archaeological remote sensing at York.' *Antiquity* 63, 337–342.

STRIGHT, M. J., 1986. 'Evaluation of archaeological site potential on the Gulf of Mexico continental shelf using high-resolution seismic data.' *Geophysics* 51, 605–622.

TABBAGH, A., 1973. 'Méthode de prospection électromagnétique S. G. D. utilisation de deux sources.' *Prospezioni Archeologiche* 7–8, 125–133.

TABBAGH, A., 1986. 'Applications and advantages of the Slingram electromagnetic method for archaeological prospecting.' *Geophysics* 51, 576–584.

TARLING, D. H., 1983. *Palaeomagnetism.* Chapman & Hall, London and New York.

TAYLOR, T. P., 1979. 'Soil mark studies near Winchester, Hampshire.' *Journal of Archaeological Science* 6, 93–100.

THOMPSON, F. H., 1979. 'Three Surrey hillforts: excavations at Anstiebury, Holmbury and Hascombe, 1972–1977.' *Antiquaries Journal* 59, 245–318.

THOMPSON, F. H. (ed.), 1980. *Archaeology and Coastal Change.* Society of Antiquaries, London.

THOMPSON, R. and OLDFIELD, F., 1986. *Environmental Magnetism.* Allen & Unwin, London.

TITE, M. S., 1966. 'Magnetic prospecting near the geomagnetic equator.' *Archaeometry* 9, 24–31.

TITE, M. S., 1972a. *Methods of Physical Examination in Archaeology.* Seminar Press, London and New York.

TITE, M. S., 1972b. 'The influence of geology on the magnetic susceptibility of soils on archaeological sites.' *Archaeometry* 14, 229–236.

TITE, M. S. and LININGTON, R. E., 1975. 'Effect of climate on the magnetic susceptibility of soils.' *Nature* 256, 565–566.

TITE, M. S. and MULLINS, C., 1970. 'Electromagnetic prospecting on archaeological sites using a soil conductivity meter.' *Archaeometry* 12, 97–104.

TITE, M. S. and MULLINS, C., 1971. 'Enhancement of the magnetic susceptibility of soils on archaeological sites.' *Archaeometry* 13, 209-219.

UHLIR, A., 1955. 'The potentials of infinite systems of sources and numerical solutions of problems in semiconductor engineering.' *Bell System Technical Journal* 34, 105-128.

WAINWRIGHT, G. J., 1979. *Mount Pleasant, Dorset: Excavations 1970-1971.* Society of Antiquaries, London.

WAINWRIGHT, G. J. and LONGWORTH, I. H., 1971. *Durrington Walls: Excavations 1966-1968.* Society of Antiquaries, London.

WATERS, G. S. and FRANCIS, P. D., 1958. 'A nuclear magnetometer.' *Journal of Scientific Instruments* 35, 88-93.

WENNER, F., 1916. 'A method of measuring earth resistivity.' *Bulletin of the U.S. Bureau of Standards* 12, 469-478.

WEYMOUTH, J. W., 1986. 'Geophysical methods of archaeological site surveying.' *Advances in Archaeological Method and Theory* 9, 311-395.

WEYMOUTH, J. W. and WOODS, W. I., 1984. 'Combined magnetic and chemical surveys of Forts Kaskaskia and de Chartres I, Illinois.' *Historical Archaeology* 18 (2), 20-37.

WILLIAMSON, T., 1987. 'A sense of direction for dowsers?' *New Scientist* 113, 40-43.

WILSON, D. R., 1982. *Air Photo Interpretation for Archaeologists.* Batsford, London.

WOODS, D. V. and KRENTZ, D. H., 1984. 'Integrated geophysical investigation of the Fort Frontenac archaeological site.' *Expanded Abstracts of the 54th International SEG Meeting*, 182-185. Society of Exploration Geophysicists, Tulsa.

YATES, G., 1989. 'Environmental magnetism applied to archaeology.' Ph.D. thesis, University of Liverpool, unpublished.

Glossary

A list of terms not fully explained in the text, or re-used remotely from their definitions.

Algorithm. A systematic mathematical procedure that enables a problem to be solved in defined steps, especially by a suitably programmed computer.

Apparent resistivity. The mean resistivity value of the ground as measured by an electrode array, including any non-uniformity of the soil and of any objects buried within it.

AC. Alternating current, electric current constantly changing direction. The number of times a complete positive and negative cycle occurs per second is called the frequency, measured in Hertz.

A/m. Amperes per metre. The unit of magnetic field strength, the intensity of magnetization per unit volume. It is the field strength induced within a loop of wire with a radius of 1 m, through which a current of 1 ampere is flowing.

Anomaly. In the geophysical context, instrument readings contrasting with the general 'background' level. Positive anomalies are above, negative anomalies below, the general level.

Barrow. A burial mound. Often marked on maps as 'tumulus'.

Berm. A platform of level ground between an earthwork and its accompanying ditch, especially between the central mound and surrounding ditch of a Bronze Age bell barrow.

BP. Years before present, convenionally defined as AD 1950.

cgs units. See SI.

Colluvium. Material, especially soil, transported downhill by a combination of weathering and gravity. Different from alluvium, which is eroded and redeposited by rivers.

Curie point. The temperature at which a ferromagnetic substance loses all its remanent magnetization.

Cursus. A long, linear prehistoric landscape feature defined by a pair of parallel ditches. The name is Latin for 'racecourse', which they could have been, but a ritual function is generally assumed.

DC. Direct current, electric current flowing in one direction only.

Dipole. Two equal magnetic poles of opposite sign separated by a very short distance, in effect a very small bar magnet.

Diurnal. Belonging to one day; daily.

Domain. The volume of a magnetic material in which all the magnetization is in one direction. Larger grains are multi-domain; single domain grains are, for instance, between 0.05 and 0.35 m (millionths of a metre) for magnetite.

Electromagnetic radiation. Energy waves created by the acceleration of electric charge associated with a combination of electric and magnetic fields.

Equipotential. A notional line or surface, all parts of which have the same electrical potential in relation to a chosen reference.

Evapotranspiration. The combination of evaporation and transpiration (taking up and exhalation by plants) by which soil loses water to the atmosphere.

Field capacity. The proportion of water retained by a soil in a condition of equilibrium: any additional water will drain away.

Free radical. A group of atoms which usually exists in combination with other atoms, but which may exist independently for short periods under special conditions.

Frequency. The speed of oscillation of a wave. Measured in Hertz.

Geomagnetic field. The Earth's magnetic field.

Grubenhaus. A German word used for a Saxon sunken hut. Sometimes colloquially called a 'grub-hut' byBritish archaeologists.

Harmonic. Production of electromagnetic waves by an instrument is often accompanied by harmonics, which have frequencies that are whole multiples of the fundamental frequency.

Heading. The direction in which an instrument is aligned relative to north.

Hertz (Hz). Oscillations (cycles) per second, for instance of an AC transmitter. Named after Heinrich Hertz.

Impedance. Restriction of the flow of electric current in an AC circuit. Resistance is a form of impedance, but it also includes the effect of inductors such as chokes.

Inverter. A device that converts a DC voltage supply into AC.

Magnetic moment. Total magnetic effect of a magnetized volume of material. It is field strength multiplied by volume: Am2.

Magnetostratigraphy. The study of deposition in terms of the magnetic properties of the layers.

Modulation. Variation of a wave of one frequency by the superimposition of a wave of a different, usually lower, frequency.

Multiplexer. A multiple switching system making and breaking a number of connections simultaneously.

Nanosecond. 10^{-9} or one billionth of a second.

nanotesla (nT). See tesla.

Natural. A term much used in archaeology for ground undisturbed by man. It can be bedrock or subsoil.

Noise. A term borrowed from acoustics to denote any incoherent instrument signal caused by random variations in the measured quantity, or in the instrument itself.

Opto-electronic. An electronic device incorporating a light and photocell arrangement. An opto-electronic distance transducer is driven by pulses of light passing through a rotating wheel with alternate opaque and transparent sectors.

Pedogenic. Soil-forming.

Pedology. The study of soil; soil characteristics.

Phase. The timing of a repeated effect, for example the negative and positive oscillations, or waves, of an AC transmitter signal. An in-phase response is in time with the transmitted waves.

Pixel. One of the basic individual elements used in constructing a picture. They may be graded in density or size, the latter as in the dots used to create a half-tone picture.

Podzolisation. The leaching out of iron and humic material from acid soils, especially in areas of high rainfall. This leaves the upper layers bleached to a light-grey colour, while the leached material tends to redeposit lower down, sometimes as a hardpan.

Polarization. Electrochemical voltages set up at interfaces between metals and electrolytes (in soils in our case), as the result of current flow. The effect also occurs with different geological materials, notably clays in the present context, forming the basis of induced polarization.

Potentiometer. An instrument for measuring electrical quantities by balancing an unknown potential difference or voltage against a known potential difference.

ppm. Parts per million.

Regression line. A line of best fit drawn through a set of points on a graph.

Remanent magnetization. Magnetization retained in the absence of a magnetizing field.

Rendzina. Rich, dark humic soil common on chalk and limestone and calcareous blown sand in north-west Europe.

Skin effect. In a conductor carrying an alternating current (AC), an electromagnetic effect which causes the current to be greater near the surface of the conductor than deeper within it. The effect increases with frequency.

SI. The International System of Units (*Système International d'Unités*), based on the metre, kilogram, second, etc. The old cgs system was based on the centimetre, gram, second, etc. The mass-specific unit SI/kg used in this book is one hundredth of the more formal unit $\mu m^3\ kg^{-1}$, but is simpler to write and better suited to the order of magnitude of soil susceptibilities and instrument scales.

Signal. The output of an instrument.

Silt. In archaeology, this term is not limited to the fine partical size range by which it is defined in more precise sciences, and is applied generally to the naturally accumulated fillings of ancient excavated features such as ditches.

Sine wave. A very 'pure' AC wave. Technically, a wave in which one variable is proportional to the sine of the other.

Sites and monuments record (SMR). Descriptions and locations of all known ancient sites and monuments in an area. In the United Kingdom, usually maintained by county planning departments, most of which have an archaeology section.

Solenoid. An electromagnet made by passing an electric current through a cylindrical multi-turn coil. The magnetic field associated with the flow of current in a wire is reinforced in this way to produce a net magnetic field parallel with the axis of the coil.

Standard deviation (s.d. or $ID). A statistical method for estimating the dispersion of a set of values around their arithmetic mean, avoiding excessive bias from extreme values. It is the root mean square of the deviations from the mean.

Tare. In weighing, the weight of the container. Also used as a verb. Most modern balances will deduct the tare automatically once the empty container has been weighed.

Telluric. Pertaining to the Earth. Telluric currents are natural electric currents.

tesla (T). Magnetic flux density, the strength of magnetic field per unit area (square metre). Named after Nikola Tesla. The tesla is a very large unit, and it is more convenient in the present context to use the sub-unit nanotesla (nT), which is 10$AB9, or one billionth of a tesla.

Thermal capacity (heat capacity). A measure of the amount of heat required to raise the temperature of a material by a certain amount. Specific heat capacity is expressed in joules per kg per Kelvin (degree C).

Transducer. In this type of context, a device for converting one physical quantity into another, for example distance into voltage.

Traverse. In the context of this book, a line of instrument readings in the field.

Uniselector. A multi-contact rotary switch driven by a solenoid.

Vicus. A civilian settlement outside a Roman fort.

VLF. Very low frequency, and therefore long wavelength, radio waves.

Water balance. The net water input (which can be negative) to soil, obtained by subtracting losses by evapotranspiration from gains by precipitation. Measured in millimetres or inches.

XY plotter. An instrument for plotting one electrical voltage against another as a graph on paper. One voltage drives the pen horizontally (the X-axis), the other vertically (the Y-axis).

I must acknowledge the value of the Penguin Dictionaries of Electronics, Geology and Science in compiling this Glossary, and I recommend them for further pursuit of definitions.

Supplement

Chapter 2 – Resistivity

Conductivity survey

Comparisons of resistivity and magnetometry with electromagnetic conductivity surveys have yet to be fully explored. Conductivity instruments are relatively easy to use, especially over hard surfaces where resistivity is impossible, and seem to be more neglected than they should be. Used in susceptibility mode, they also have some attractions compared with magnetometers (Tabbagh, 1984). The Geonics EM38, an instrument capable of both conductivity and magnetic susceptibility surveying, and particularly suited to work on the archaeological scale, was introduced in Chapter 2. An example of its potential is shown in Fig. 120. Bevan (1993) lists the disadvantages of this class of instrument as lack of user control of detection depth, too great a sensitivity to metal, and, in the case of the EM38 at least, thermal drift, especially in and out of sunlight, requiring the addition of insulation. In the face of these problems, the attractions of rock solid resistivity readings are understandable. However, Bevan has since carried out a thorough evaluation of the EM38 as part of geophysical tests on Jamestown Island on behalf of the Colonial Williamsburg Foundation. He found that its height sensitivity was not as great as many suspected: raising the instrument from 5 cm to 35 cm (2 in to 14 in) above the ground decreased the signal by only 17 per cent. Response can be maximised and made more consistent by mounting the instrument on a sledge (Fig. 121), which also facilitates its use horizontally if shallower penetration, more closely comparable to resistivity, is required (cf. Fig. 25). In susceptibility mode, the best comparison with magnetometry seems to occur with the EM38 vertical (Neil Linford, pers. comm.).

In a comparison with 1 m Wenner resistivity, Bevan showed the dynamic range of the EM38 (disallowing a large number of high readings probably due to metal contamination) to be relatively cramped. A range of 8.5–13.5 millisiemens/m (equivalent to 70–120 Ω-m), considerably less than predicted by theory, coincided with a resistivity range of about 75–260 Ω-m. This discrepancy could be explained by the argument on p. 36 of this book.

Electrical imaging

Comprehensive and rapid three-dimensional imaging of the ground becomes more achievable as data collection and logging become more sophisticated. With modern equipment it is increasingly easy to collect large datasets for pseudo-sections or tomography (p. 61). Griffiths and his colleagues have developed the Campus Geopulse for automatically collecting the necessary data by way of a multi-contact cable stretched across the ground, after which the data are processed by means of a laptop computer in the field to produce cross-sectional resistivity contour plots (Griffiths and Barker, 1994).

Walker has introduced the Geoscan RM15 instrument which is more powerful and flexible than the RM4. Combined with the PA5 multi-probe array, an extension of the twin-electrode frame with multiple electrodes, it makes possible the plotting of a survey as a series of stacked pseudo-sections – a substantial advance in information content and presentation, although it should be repeated that the resistivity profile of a feature may not coincide with its physical profile, as discussed in Chapter 2.

Allied to electrical imaging is the much more basic approach of simply using two probe spacings to obtain normal plots or profiles at different depths (a rather different use of such data from that proposed on pp. 155–6). In this way, a useful dual-depth survey was achieved at Reigate Priory (see Fig. 120). Herbich (1993a; 1993b) favours the use of two different spreads of the Schlumberger array. With this combination, he has been very successful in defining the areas of Polish flint mining sites, where the fairly uniform mixing of upper geological layers by the mining activity produces similar resistivity values at the two penetration depths, compared with strongly contrasting values where the geological layering is undisturbed. He has also used the method successfully to distinguish structures of different periods on a Roman town site (Herbich et al, 1993)

Resistivity in site analysis

The use of resistivity search traverses has been briefly noticed above (p. 46), the assumption being that one is looking for buildings, roads or perhaps linear ditches. Carr (1982) has taken a substantial

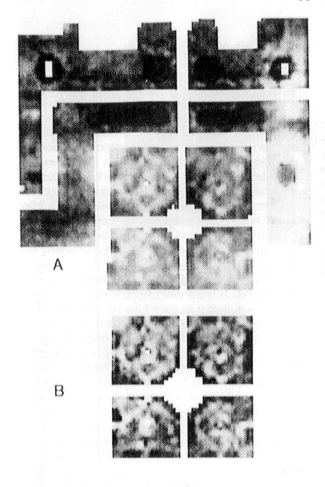

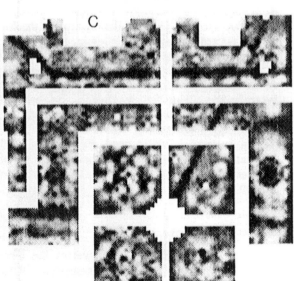

Fig. 120. Reigate Priory gardens: part of a survey and comparative exercise by the Ancient Monuments Laboratory. The present house is at the top (north side) of the plots. Readings were taken at 1 m intervals, and plotted with high values light. A: 0.5 m twin-electrode survey in which the clearest feature is a geometrical former garden layout, each 'a uniform diamond with marked spurs on each corner'. This probably dates to the nineteenth century. The ability of resistivity to define such supposedly transient features – which must survive as little more than variations in soil texture – is proving to be of great value in the developing study of garden archaeology. B: Detail of the same area with contrast enhancement. C: Contrast enhanced 1 m twin-electrode survey of the same area. At this spacing the detailed garden pattern breaks up, but deeper features now appear, such as the diagonal line of a water supply running to its centre, and another likely pipe along the front of the house. In another part of the survey, not illustrated, possible building foundations appeared which were not visible at 0.5 m spacing. D: EM38 conductivity survey of part of the area. There is a good match with the 1 m resistivity, but the low resistance pipes now appear as strongly conductive positive anomalies. An advantage of this technique is that the survey can be continued over the paths; a disadvantage is that it is affected by positive and negative noise caused by metal rubbish, which gives the survey a 'speckled' appearance. Survey report by Neil Linford, English Heritage.

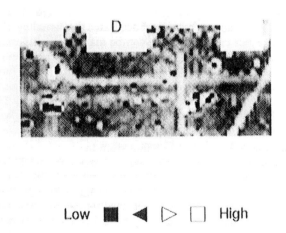

Low ■ ◀ ▷ ☐ High

0

Fig. 121. Lewis Somers using a sledge-mounted EM38 at Avebury. The instrument is fixed vertically for maximum depth penetration. A Hewlett Packard hand-held computer is used as a data logger.

step beyond this to suggest that in intra-site studies resistivity may be capable of characterising and distinguishing former activity areas. He examines in detail the different types of chemical input likely to be produced by different activities, and the formation of humus, how these affect the resistivity of the soil and how their effects are modified by climatic variation and time. The variables seem rather daunting, but some encouraging results were reported. The subtle texture contrasts recorded in garden surveys (see Fig. 120) are a similar phenomenon.

Electrode configurations

Interest in a configuration known as the pole-pole has developed since the first edition of this book was published. It is similar to the twin electrode, except that the fixed probes are placed widely apart so that their spacing ceases to be critical. The setting out process is described by Bevan (1993): 'Typically, I will set the current reference electrode about 200 m (600 ft) distant from the middle of the survey area. Then the voltage reference electrode is taken about 100 m (300 ft) toward that electrode and then about 100 m (300 ft) to the left or right; this keeps it at about the zero potential line between the fixed and moving current electrodes.' The resistivity formula at the moving probes then reduces to the same as for Wenner, so that there is no need for adjustment when moving the fixed probes. The only problem so far reported is an increased sensitivity to interference because of the long cables. In his 1993 publication, Bevan also includes a useful discussion of resistivity instruments and other techniques.

Electrostatic electrode systems

It has recently been appreciated that resistivity measurements can be made on the ground with non-contacting electrostatic electrodes which induce a voltage in the ground without direct coupling (Grard and Tabbagh, 1991; Tabbagh, 1993). This remarkable development was stimulated by the use of such electrodes for measurement of the resistivity and permittivity of space plasma on Earth-orbiting satellites.

The prototype ground-based system is a rectangular array 1.17 m × 1 m (3.84 ft × 3.28 ft), supported on two wheels. The electrodes are short lengths of heavy metal chain wrapped in plastic bags, the weight of which keeps them as close to the ground as possible when they are dragged along. The injection and receiver circuits have individual power supplies, phase-synchronised by means of an opto-electronic coupler. The receiver (potential measuring) electrodes have preamplifiers with unity gain. Current into the transmitting electrodes is about 5mA, and the operating frequency is 128 kHz.

The accuracy of a 1.5 m (4.92 ft) square version of the apparatus is negligibly affected by a change in height of the electrodes from ground level to 7 cm (2.8 in). Even so, it would probably not be successful even on quite moderately vegetated ground, but it is suggested that wheel or brush contacts would keep better contact. In a test survey, results compared closely with a conventional resistivity survey, and the system was demonstrated to be effective for defining features under the floor of the Cathedral of Autun, even where this was overlain by a steel-reinforced concrete slab (thickness not quoted). It seems likely to be a potent rival to ground penetrating radar in such situations.

Resistivity theory

The following analyses (Aitken, 1961, 1974; Aspinall & Lynam, 1970) established the essential equations of earth resistivity measurement.

Single electrode. Consider a single electrode probe making point contact with the surface of ground of uniform resistivity. A current I maintained by a battery with a remore earth-return flows from the probe in straight radial lines and the equipotentials are hemispherical surfaces so that at a radius r the current crossing unit area of such a surface is given by j, where

$$j = \frac{r}{2\pi r^2} \qquad (1)$$

Since j is related to the electric field E at any point by the relation

$$j = \frac{E}{\rho} \qquad (2)$$

it follows that

$$E = \frac{\rho I}{2\pi r^2} \qquad (3)$$

where ρ is the resistivity. By integration, the potential at any radius r is given by

$$V = \frac{\rho I}{2\pi r} \qquad (4)$$

This is the fundamental equation of earth resistivity measurement.

General four-electrode array In equation (4), substituting R, the measured resistance, for V/I gives

$$R = \frac{\rho}{2\pi r} \qquad (5)$$

In the general four-electrode array depicted in Fig. 122, substitute for r in terms of a and b in equation (5) to obtain the net potential difference between P_1 and P_2 due to the sources $+C_1$ and $-C_2$:

$$R = \frac{\rho}{2\pi}\left[\frac{1}{a} - \frac{1}{Na} + \frac{1}{b} - \frac{1}{Na - a + b}\right] \qquad (6)$$

If we make the condition b = a, then (6) becomes

$$R = \frac{\rho}{2\pi}\left[\frac{2}{a} - \frac{2}{Na}\right] \qquad (7)$$

Rearranging gives

$$R = \frac{\rho}{\pi a}\left[\frac{N-1}{N}\right] \qquad (8)$$

or

$$\rho = \left[\frac{N}{N-1}\right]\pi a R \qquad (9)$$

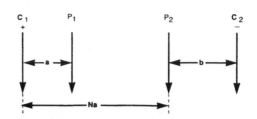

Fig. 122. General four-electrode array.

These enable us to obtain expressions for ρ in terms of R, a and b for the various configurations of probes. The simplification when b = a, evident from equations (8) and (9), is relevant to the configurations found most successful in this book. For instance, for the Wenner array, $\rho = 2\pi a R$; Double Dipole, $\rho = 6\pi a R$; Square, $\rho = \frac{2}{2 - \sqrt{2}}\pi a R$; Twin (assuming both pairs of probes are in uniform ground of the same resistivity), $\rho = \pi a R$. The strength of this configuration lies in the fact that, as N becomes increasingly large, the ratio of N to N − 1 tends to 1, so that the background reading level changes little as N is varied; thus it does not change more than 3% when the separation is increased from 30a to 300a, and a minimum separation of 60a will reduce this factor to a mere 1.3%.

Chapter 3 — Magnetometry

Magnetometers

Of the two types of magnetometer discussed below, the Overhauser is new to this book, while the caesium has been described; both share some similarities of functional principle (Scollar *et al.*, 1990). These instruments should be regarded as examples of their type, but both are produced by companies with an active interest in their archaeological application, and provide an opportunity to examine some practical aspects.

Overhauser magnetometers. This type of instrument, developed from the concept of the proton magnetometer, has become increasingly available since the first edition. The GSM 19 instrument of GEM Systems is available in absolute or gradiometer configuration with simultaneous measurement, adaptable as a straight proton instrument if required, and even as a VLF instrument. It makes use of a mixture of proton and electron-rich liquids, with the addition of free radicals, which in combination strengthen the signal. Polarisation, rather than by the 'sledgehammer' of a DC current, is achieved with a radio frequency alternating voltage automatically tuned to follow the frequency of the proton/electron precession. Thus the precession is continuously refreshed, and the instrument can give up to five readings per second. Assuming a walking pace of 1 m in 0.7 second (about 5 ft per second), this is equivalent to 3.5 readings per metre (1 reading per foot), which is just fast enough to give the level of spatial resolution required in archaeology. Sensitivity resolution is 0.01nT. In gradiometer form, a built-in data logger stores 6800 readings as standard, with the option of 110,000.

Caesium magnetometers. The principle of these is described in the main text. The new Geometrics G-858, available in gradiometer form, is functionally typical, and has attractive new display features. Cycle-rates of up to 0.1 second ensure that fully adequate spatial resolution is achieved, although at some expense of usable sensitivity, so that at a typical 0.2-second cycle-rate this is reduced to 0.03nT. Accurate operation depends on keeping the sensor orientation within $\pm 35°$. Battery life of the gradiometer is at present restricted to three hours.

An attraction of this instrument is that it has a large LCD display interactive with data logging facilities. The most valuable attribute of this is its ability to display the last five lines of data in graphical form, a boon to the understanding of magnetic scans. Additional software makes possible the immediate production of good-quality survey maps, and dipole pattern matching software will produce maximum likelihood estimates of object size, depth and possible identification.

Instrument summary. Bearing in mind background noise due to electronics consoles, batteries and other causes, and the balance of advantages and disadvantages, it seems that the major instrument types described, including fluxgate instruments, when configured as gradiometers are, in practice, of comparable sensitivity and effectiveness. Fluxgate gradiometer surveys can be usefully plotted over a range as narrow as 1nT from white to black. The two types of instrument described here, however, gain significantly in useful sensitivity when configured as differential instruments with remote reference sensors (see Fig. 59), although there is then the disadvantage of trailing leads. Overall, the fluxgate gradiometer retains the advantages of compactness, ease of use and price, but would benefit from some of the emerging additional features of the other instruments. Its main drawback is the need for careful setting up, although this may soon be overcome.

More extended discussions of instruments are given in Bevan (1991a) and Scollar *et al.* (1990), the latter including thorough theoretical treatment.

Chapter 4 – Magnetic susceptibility

Large evaluation surveys

The increased need for large-scale archaeological evaluation surveys as part of the planning process has been introduced in the Preface. The subject is admirably and extensively addressed in English Heritage Research and Professional Guideline No. 1 (David, 1995). The prime aim here is to update this aspect of the present book, placing it in the supplement to the chapter on magnetic susceptibility in recognition of the contribution this technique is now making to the subject.

Mainly because of their speed and specificity of response to the soil-altering effects of past human settlement and industrial activity, magnetic techniques have emerged as most suitable for broadscale work. Before the recent increase in the use of magnetic susceptibility, large areas were almost invariably examined by means of detailed fluxgate gradiometer survey, preceded or supplemented by magnetometer scanning as appropriate (pp. 83–91). Total detailed coverage was generally only economical over limited areas or where archaeology was known to be dense (e.g. Fig. 107). Scanning as a preliminary approach is now joined by magnetic susceptibility – which may also provide the only usable magnetic evidence where interference is serious.

Capabilities of magnetic susceptibility in evaluations (Figs 123, 125). Around most archaeological settlement and industrial sites there is a spread of detritus-affected soil, or discard area, that can sometimes be seen as a dark halo from the air and is detectable on the ground by its enhanced magnetic susceptibility. Because of this, with assis-

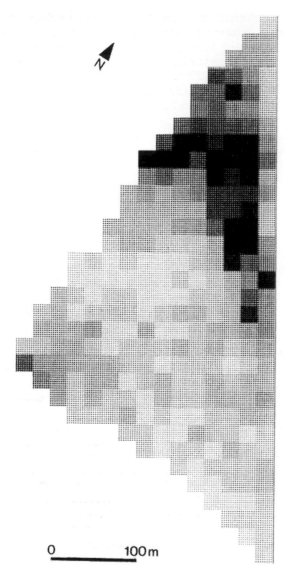

0 ⟼ 100 m

Fig. 123. Magnetic susceptibility in good conditions. The site is on alluvium derived from the immediately adjacent iron-rich soils of the Jurassic Ridge in Northamptonshire. The Roman Watling Street runs just outside the survey area along the north-east boundary of the field. (Left) The MS2D field loop survey at 20 m intervals, plotted over the range 0 (white) to 40×10^{-5} SI (black) volume susceptibility. The general level of the background is about 10×10^{-5} SI. (Right) The confirmatory fluxgate gradiometer survey, plotted over the range -2nT (white) to +2nT (black) shows circular outlines, probably of houses, and other features within enclosure ditches; overlying ridge and furrow can also be seen. Surveys by Stratascan for RPS Clouston.

tance from subsequent ploughing, even a small site can affect the soil over quite a large area, and is confidently recognisable against a background of uniform susceptibility as a characteristic 'smudge', tapering away at the edges, on a half-tone or dot density plot. This confidence tends to be reduced in the case of linear surveys, where the visibility of the context is constrained by the narrowness of the survey strip. Sharply high readings in ones or twos, without the tapering effect, are likely to be of relatively modern origin, and tend to occur along existing field boundaries and roads, and behind back gardens (Fig. 124). One must also be wary of small-scale geological patterns. Measurements around the experimental earth-work on Overton Down, Wiltshire, to be reported in a forthcoming Council for British Archaeology monograph, suggest the presence of relict 'stripes' of increased susceptibility values coinciding with increased clay content of the normal chalk rendzina soil. Such stripes can be only a few metres in width and are probably periglacial in origin.

Susceptibility surveys are effective with exceptionally wide and therefore economical sampling intervals. In search mode, as discussed in the main text (and see Fig. 105), even more widely spaced grid intervals or point readings have proved successful, but the coarsest recommended for routine work is now 20 m (say 60 ft), at which one can expect several readings to be affected by a medium-sized site and its spread of material. If resources allow, a 10 m (30 ft) interval is preferable: even though it quadruples the work involved, it reduces the chance of missing lesser sites, and is capable of resolving some features of past landscape use, for instance the lines of old boundaries showing as disturbed values, which can be caused by dumping or where ditch digging has brought relatively low susceptibility soil to the surface.

Narrower intervals are generally too labour-intensive for extensive search work. In a limited study area, a 5 m (15 ft) interval can be useful, while intervals less than this are better reserved for planning individual features in detailed site work where conditions are suitable (e.g. Fig. 86). Such use on excavations will also be discussed in the next section.

As described in the main chapter, magnetic susceptibility has a particular contribution to make over igneous geology. It is also responsive to subtle sites without cut features, perhaps visibly represented by no more than a few chipped flints, and undetectable by magnetometers.

Finally, magnetic susceptibility is now so relatively undemanding and economical compared with other methods that some workers include it automatically with other types of survey for the added support it can provide in interpretation.

Limitations of magnetic susceptibility. Because of their shallow penetration, magnetic susceptibility instruments are particularly sensitive to the occluding effects of deposition and soil movement, and measurements will not be effective if the old surface is overlain by later accumulations of colluvium, tephra, peat or alluvium – at least when more than about 0.5 m (1.7 ft) deep (see main chapter and Clark, 1992). These problems can sometimes be overcome by using a borehole probe, although the method then loses its advantage of speed. Problems may also be caused by geological variations masking those of archaeological origin (see below), the translocation of soil by plough, gravity and erosion, and alteration and movement of iron oxides by gleying and podzolisation. An appreciation of geomorphology, topography, past land exploitation and history can be helpful in anticipating these effects. It must also be emphasised that most of the assumptions and experience described here relate to a landscape largely of farmland on the soils and in the temperate climate of Britain. The suitability of an area for magnetic susceptibility surveying should always be tested – for instance, J. Orbons (pers. comm.) has had limited success in tests on Dutch soils – and it should normally be used in combination with other techniques.

Problems can also be encountered with archaeological features that contribute little or no magnetic enhancement to the soil compared with occupation and industrial activity. Characteristic of these are cross-country ditches and ceremonial sites, especially those of prehistoric date, such as barrows, henge monuments and cursus. However, the ditches and other cut features of such sites are normally detectable by magnetometer survey, and it is standard practice to supplement susceptibility with a gradiometer scan, which will be facilitated by the development of instruments producing continuous graphical plots on a built-in screen.

Types of evaluation survey and their implementation

At present, the best combination of economy and reliability is probably achievable by full magnetic susceptibility coverage and detailed magnetometer sampling of about half the area. A large area can be surveyed by magnetometer in alternate strips 20 m (60 ft) wide, or preferably 30 m (100 ft) wide to reduce the work of surveying in. Ideally, this would be done in a chequerboard pattern of squares to avoid missing any lineations between

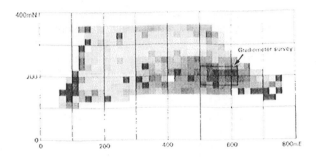

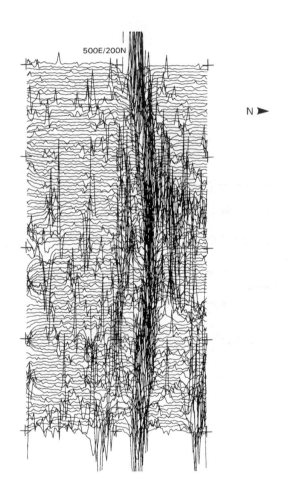

Fig. 124. (Top) Magnetic susceptibility survey of the former Hurst Park racecourse, on sand and gravel terraces of the River Thames in Surrey; a 'classic' magnetic susceptibility site with good subsequent excavation control. Because of rough vegetation and iron contamination, readings were obtained from topsoil samples taken at 20 m intervals. They ranged from about 10 to 80×10^{-8} SI/kg and are plotted between the mean -1 (white) to +1 (black) standard deviations ($\pm 36 \times 10^{-8}$ SI/kg) by means of Geoplot 2.01. Individual extreme readings are mostly associated with 'discard areas' of modern houses along the western edge of the survey area, and with demolished racecourse buildings at the south-east corner. The substantial area of high readings represents a multi-period site. (Below) The rectangular area containing the highest of these readings was surveyed by fluxgate gradiometer (the plot shows raw data at 6nT/traverse interval) but interference caused by pipes and other magnetic objects was so dominant that any response to archaeological features was obscured. Therefore the whole survey depended upon the susceptibility measurements. A very low correlation of $r = -0.19$ for 19 samples across the enhanced area supports the anthropogenic interpretation of the high values.

There was some complaint from the archaeologists that the high susceptibility readings had too westerly a spread. Two factors probably explain this: heating experiments showed a peak in the natural iron content of the soil at this end of the enhanced area, and hence some geological variation; also there is a step change in readings along the grid line 300mE, which approximates closely to a parish boundary. Early maps show a complex of fields east of this boundary, but much less agricultural activity to its west. Thus it seems that cultivation has tended to spread the high-susceptibility soil up to the boundary. A small (*c.* 3 per cent) increase in frequency dependence under the archaeological peak indicated that the possible loss of fine grains might not have been as great as at Lavendon. This, combined with the natural peak, probably accounts for the lack of a clearer negative correlation between samples before and after heating.

A band of low readings centred on the grid line 300mN seems to represent a palaeochannel containing an alluvial silt probably also affected by gleying.

The report on the excavation of this site will be published in a monograph by Wessex Archaeology. The survey was commissioned by Cotswold Archaeological Trust acting for Wates Built Homes Ltd.

the strips, but would involve a substantial increase in the complexity of marking out and actual survey work. A linear strip such as a road route can usually be sampled satisfactorily by means of a continuous magnetometer survey 20 m (60 ft) wide along its centre, supplemented by a similarly centred susceptibility survey at 10 m (30 ft) reading interval and four readings wide, covering in effect a strip of 40 m (120 ft) in width.

Gaffney and Gater (1993) propose three levels of geophysical investigation of increasing detail: Level I (prospection), Level II (assessment) and Level III (investigation). The schemes proposed below largely represent different approaches to their

Level I. These schemes are also placed in order of thoroughness and detail, but the word level has been avoided to prevent confusion.

Scheme 1 is the most thorough, while scheme 2 is more satisfactory than scheme 3 because of the inclusion of unbiased magnetometer area coverage. Where resistivity is bracketed, it is regarded as an optional refinement for sites already identified by the other methods and suspected of containing buildings.

Geophysical evaluations are often followed up by more-or-less random machine trenching, except where the existence of sensitive features has been clearly established by one means or another. While fortunately not the responsibility of the geophysicist, these should be examined if possible for the interpretative feedback they provide.

Scheme 1

Full magnetometer survey
Magnetic susceptibility on 10 m (30 ft) grid
Resistivity over likely buildings

Scheme 2

Magnetic susceptibility on 10-20 m (30–60 ft) grid
Magnetometer scan
Full magnetometer survey of arbitrarily determined
    sample areas – preferably at least half the total
    area and if necessary intensified over areas
    suspected of significance on the basis of air
    photographs/fieldwalking/sites and monuments
    record (SMR) or other documentary research
Minimum 20 m (60 ft) strip for linear projects
(Resistivity over likely buildings)

Scheme 3

Magnetic susceptibility on 10–20 m (30–60 ft) grid
Magnetometer scan
Full magnetometer survey of areas defined by above
    methods and by air photographs/fieldwalking/
    SMR or other documentary research
(Resistivity over likely buildings)

Practicalities of large-scale magnetic susceptibility surveys

Sampling methods. As previously discussed, the standard sensor for field measurements is the Bartington MS2D loop. A small stand-off leg or spigot (MS2DS), recently introduced and also avail-

able as an add-on to existing loops, can be used to maintain the sensor at constant height over a considerable range of surface conditions and vegetation cover. This greatly improves consistency of response, though at the expense of sensitivity. If circumstances arise where it is inadequate, the MS2F field probe or MS2G downhole sensor may be used in combination with the Eijkelkamp grass plot sampler, type 05.10, in the manner of Fig. 82. This is also excellent for taking samples for laboratory measurement. It consists of a slightly tapered tube attached to the bottom of a cup with a long handle, all made of stainless steel. The tube is pushed into the ground by foot pressure on the cup, and the second insertion pushes the first sample into the cup. In very wet clayey conditions, one can resort to taking samples with a gouge auger which indeed is favoured by some for all sampling. With such crops as ridged-up potatoes, it can be possible to use the MS2D against the ridge sides but, as I write this, a colleague reports that he has to survey a potato field where the plants are so thick that it is only possible to take samples, although conditions are very dry and these can be readily scooped from the surface. Samples are most conveniently collected and measured in the MS2B in the plastic pots used for 35 mm film, which can be obtained free from photographic processing shops, where they are normally discarded – transparent pots have one advantage (see below), otherwise the Kodak version is generally the best (and in Britain the Boots version), and most consistent in tare weight.

Field logistics. It is assumed that one of the above methods of field measurement will be used, deployed as described for sampling surveys on p. 163. If samples are being collected, it is best to use pots prenumbered in a standard order, the pots being used directly for measurement in the MS2B, and reserved as required for further treatment.

Artificial or natural? – tests for magnetic susceptibility anomalies

Changes in soil susceptibility can have both archaeological and geological causes, and it is crucial to be able to distinguish between these. An obvious approach is to test all enhanced areas with detailed magnetometer survey, but this may be prohibitively demanding in resources, or may not be conclusive if a site has few cut features or there is bad magnetic interference. An alternative is to apply methods using intrinsic properties of the soil.

'Maximum conversion' is a laboratory heat treatment which converts all available iron compounds in the soil to secondary ferrimagnetic oxides, apparently dominated by maghaemite, which are those produced by burning and other effects associated with human occupation. 'Fractional conversion', the ratio of the original susceptibility of the soil to that after maximum conversion, expressed as a percentage, is used as an indicator of the presence and strength of the human effect. It has been demonstrated that fractional conversion can be low and fairly uniform except where enhancement has occurred as a result of human occupation. In some situations, however, this assumption has proved to be unreliable because the natural level of conversion can vary with changes in underlying geology. This has prompted research into alternative ways of using the data which are discussed below.

Sample treatment. Mass susceptibility measurements on selected samples with the MS2B are combined with a heating procedure involving a reduction-oxidation cycle (adapted from Tite and Mullins, 1971) to produce the maximum conversion. This procedure cannot be claimed to be definitive, but has proved to be effective even with quite humble equipment.

Samples from the field in their film pots must first be dried. This is most conveniently done in an oven at about 30°C, or by leaving them open on a central heating radiator or a sunny windowsill for a few days. If transparent pots are used, condensation can be seen until they are dry. For quicker results, the samples can be spread out on small dishes and left for 12–24 hours – stronger heating is best avoided. They are then crushed with a pestle and mortar, separated from stones by being passed through a 2 mm sieve, and returned to their original pots. By this time the samples should be reduced to roughly 15-20 g and about half-fill the pots (the height of the MS2B sample support should be adjusted so that the samples are roughly central in the coil). It is advisable previously to measure the tare of each pot, although the Kodak and Boots pots vary by only about 2 per cent, which is not significant. The samples are weighed to within 0.1 g, and measured on the high sensitivity range of the MS2B at both high and low frequency.

The next stage is the heating process, for which an economical Carbolite ELF 10/6A furnace has proved suitable. Samples, reduced to about 10 g to ensure good penetration by the atmosphere, are transferred to 25 ml porcelain crucibles, pencilled with the sample number on the base. 1 g of plain flour is added and well stirred into each sample, to ensure initial reducing conditions. Any of the original sample left will be useful in case of a subsequent accident.

Samples are placed in the furnace, covered with lids, and the furnace chimney is closed. They are heated at 650°C for one hour after reaching this temperature which takes about 20 minutes with the Carbolite furnace, and is assumed to have occurred when the 'overshoot' indicator diode first lights up. At the end of the hour, the furnace is switched off, and in a further 20 minutes it should be cool enough for the door to be opened and the lids to be removed with tongs in readiness for the oxidation stage. The furnace is then run for another 45 minutes with the chimney open, the warm-up this time taking about 10 minutes. When the furnace has been switched off, the door is briefly opened to ensure that it contains an air atmosphere, and it is left to cool down. This is conveniently done overnight, but in a 'mass-production' situation, the door may be left open and the samples removed with the tongs as soon as the temperature permits.

The cooled samples are transferred to fresh tared film pots (after stirring with a plastic or wooden spatula to randomise any remanent magnetisation acquired) and measured in the same way as the original samples, except that they will now be so strongly magnetised that the fast low sensitivity range of the MS2 is perfectly suitable. Whatever they looked like originally, the samples will be strongly reddened by the heat treatment, an impressive reminder of the ubiquity of iron compounds and of environmental magnetism.

Using the heating data. The classic use of these data is to calculate the fractional conversion to maghaemite, as described above, on the assumption that this reaches a maximum where conversion has been artificially enhanced by human occupation or industry. This is a reliable assumption over uniform geology, but not otherwise.

The Lavendon by-pass survey, over pasture and ploughland, was an early example of the problems caused by mixed geology and stimulated the search for possible solutions. The 4 km route was initially surveyed at 15 m spacing with the MS2D loop. Fig. 125 shows the results of measurements on 13 samples collected along just over 2 km of the route – a coarse but adequate rate of sampling for comparing archaeological and geological effects. Let us see what can be deduced from the graphs.

χ_{LF} shows a modest peak to the left and a much more substantial one to the right. There was good control information on this stretch: the left-hand peak was known to represent a real Romano-British site, while there was no evidence for archaeology coinciding with the right-hand peak – certainly not a site a kilometre long.

One would have expected the $\%\chi$fd and fractional conversion to give distinctive peaks over the archaeological site, whereas they are in fact similar for the two peaks – sample 3, taken from the most intensively occupied part of the site, is only marginally the highest. However, the geological map shows that the right-hand peak coincides with Blisworth Limestone bedrock, over which a more magnetic soil has clearly developed. The geology of the rest of the stretch is mostly glacial head overlying Kellaways Beds, another type of limestone associated with the low susceptibility Oxford Clay.

Two possible solutions emerge from scrutiny of the curves. First, using a simple analysis, one observes as expected that there is a rise in χ_{Max}

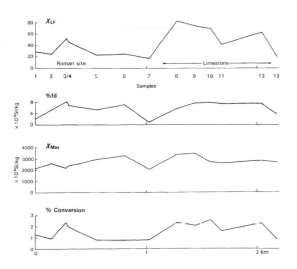

Fig. 125. Lavendon by-pass route: magnetic susceptibility. Units are 10^{-8} SI/kg.

(maximum susceptibility after laboratory heating) readings under the natural peak, but not under the archaeological peak, which is independent of the natural magnetic state of the soil.

A more formal approach is to compare the correlation between the initial susceptibility, χ_{LF}, and χ_{Max}. The natural peak should give a positive correlation. Samples 7–13 at Lavendon gave a correlation coefficient of r = 0.85 (r is the product moment coefficient; perfect correlation = 1 (positive) or –1 (negative)). On the archaeological site, however, one would expect variations in proportionality because of varying intensity of activity and burning over the site, which should give a correlation close to zero. In fact, the figures from the archaeological site, although too few to be statistically valid, hint at a negative correlation (r = –0.94), which has been subsequently found on a number of sites with larger sample sets.

The negative correlation may be due to the loss of a proportion of the finer maghaemite particles, possibly by wind or downward migration. This would tend to reduce the apparent total conversion figure, especially where the anthropogenic conversion was most strongly positive. Such movement might also be responsible for the lowness of the archaeological peak. This explanation is supported by the insignificant rise in χ_{HF}, suggesting that fine particles have been lost. Evidence of the downward translocation of such particles has been actually observed in archaeological sections by Allen and Macphail (1987).

There remains a possibility that the assumptions made above might be upset by the importation of high susceptibility material onto a site. The effect of this would be to increase both the initial susceptibilities and those after heating, giving the effect of a natural increase. In practice, no such problem has so far been encountered, presumably because imported material tends to be in large, visibly red, particles which give warning of their presence and are mostly removed during the sieving process; they are also probably dominated by the comparably strong magnetisation of the whole soil sample.

Summary of heating tests for human occupation.

The correlation method, in spite of its attractiveness, has been given second place because its reliability has not been sufficiently tested at the time of writing. It also ideally requires more samples.

Method 1

Recommended minimum of two samples on site and two samples off site, the latter preferably on opposite sides of it.

If maximum conversion shows similar results from all, then the natural is uniform, and high initial readings are likely to be due to human activity. If maximum conversion results are more than marginally higher where initial readings are high, then the effect is probably natural.

Method 2

Recommended minimum of six samples scattered over enhanced area, up to at least one of its limits.

Correlation calculated between magnetic susceptibility before and after maximum conversion. Positive correlation indicates geology; zero to negative correlation indicates human involvement.

Method 3: Areas of known uniform geology

Archaeology should be reliably distinguishable by relatively high fractional conversion – although this rarely exceeds a low single-figure percentage.

With all methods a coincident peak in $\%\chi$fd provides supporting evidence, but does not always occur. Collection of a few extra reserve samples is recommended for all methods.

Barker and Crowther (1995) have plotted laboratory-determined magnetic susceptibilities of topsoil against subsoil for a survey area, obtaining a regression line with a correlation coefficient of r = 0.722. Certain points with exceptionally high topsoil values, as well as high fractional conversions, and falling well above the regression line, were regarded as the most likely candidates for human occupation. Although the area is known to have been occupied, detailed checks had not been possible at the time of writing.

Magnetic susceptibility in excavation

Many features such as post-holes, ditches, gullies and beam slots can be traced on trench floors where they cannot be seen because of the notorious effects of drying out and the spread of dust – or simply lack of colour contrast under any conditions. Suitable instrument configurations are the Bartington MS2D loop and the WSL-B hand-held meter, made by STG of China and Japan.

After machine clearance of topsoil from two round barrows on the Northborough by-pass route, susceptibility surveys gave forewarning of burial pits (French, 1994). The technique, with a 1 m (3 ft) reading interval, was also used in preparation for the excavation of a round barrow on gravel (DEN 28) at Deeping St Nicholas, where a central burial pit and intermittent indications of other underlying features were just detectable. The burial pit lay beneath 0.5 m of overlying mound, supporting the suggestion in the main text that biological dispersion may make it possible to detect features under about this depth of overburden. Yates (1989) has also had success with this type of application.

Soil magnetic susceptibility map

For the compilation of the National Soil Inventory of England and Wales, topsoil samples were collected by the Soil Survey and Land Research Centre at each 10 km National Grid intersection across the whole of England. Analyses of various geochemical and environmental properties are included in *The Soil Geochemical Atlas of England and Wales* (McGarth and Loveland, 1992).

The samples have recently been measured at Coventry University, using the Bartington MS2B sensor, to produce the first complete maps of low frequency and frequency-dependent magnetic susceptibility for the topsoils of England. The work was undertaken by K. L. Hay, J. Dearing and S. Baban (Coventry University) and P. Loveland (Cranfield University), and is shortly to be published.

Bacterial magnetism

Fassbinder and his colleagues have reported the discovery of magnetic bacteria in ordinary meadow soil in Bavaria, and in exceptional concentration in association with the rotted wood in archaeological post holes (Fassbinder *et al*, 1990; Fassbinder, 1994). The bacteria contain magnetite, and would seem to provide the first direct evidence of the 'fermentation effect' in soil magnetism.

Chapter 5 – Other methods

Ground penetrating radar

A notable development in ground penetrating radar has been the reduction in the size of the equipment. Until recently, this has tended to fill most of a Land Rover-type vehicle and has required a separate petrol-driven generator. A contrasting example is the pulseEKKO range of instruments, produced by Sensors and Software Inc. These are battery-driven and fit into a modest-sized box. Transmitter and receiver can be manually positioned along a tape, so that the radargrams are accurately scaled for distance whatever the ground conditions (Fig. 126).

Typical frequencies used in archaeology are about 100 to 500MHz; with increasing frequency, resolution improves at the cost of penetration. In general, radar produces better images of convex or low-relief features (see Figs 92, 127) which are freer of confusing lateral reflections than steep-sided hollows or targets tending toward points or narrow lines, which give characteristic hyperbola-shaped reflections.

Radar remains most effective over uniform deposits, preferably dry for maximum penetration. Beautiful images such as that in Fig. 127 remain rare. In Britain, the promise of the work in York (p. 119) has not been sustained, but useful results have been obtained by Stratascan in the location of substantial features such as gun emplacements and vaults; and Yasushi Tanaka has been able to locate toft boundary features at the deserted medieval village at Swavesey on the edge of the Fens, using a pulseEKKO IV in extremely dry summer conditions that made resistivity very difficult.

In the United States, Bevan, a pioneer in the use of ground penetrating radar in archaeology (Bevan and Kenyon, 1975), has found radar, backed up by a soil conductivity meter (EM31 or EM38), to be the most effective way of locating historic graves, providing the geology is fairly uniform (Bevan, 1991b). The asymmetry of graves, revealed by orthogonal surveys, supported the interpretation. Graves with substantial coffins were easiest to detect, while those containing only reburied bones were not detectable. Tanaka (1995), prospecting in the alluvial soils of the Chang Jiang (River Yangtse) valley for substantial Chinese tombs, concluded that radar combined with resistivity, used as a search tool and for pseudo-sectioning, were most effective, while the EM38 in conductivity mode

Fig. 126. PulseEKKO IV: Yasushi Tanaka moving the transmitter and receiver antennae along the measuring tape. An assistant operates the control console in the distance.

was useful for finding and planning smaller and shallower features. Remarkably, the EKKO IV could be used in the most adverse conditions, ranging from swamps to house interiors, as it was almost impervious to ground conditions. A later EKKO version seems also to have been successful in detecting subtle remains through a 1.5 m (5 ft) concrete foundation in the basement of a building in London.

Improvements in speed, convenience and positional accuracy of ground penetrating radar are encouraging the use of area surveys. As with magnetometry and resistivity, the horizontal information thus obtained can be a great help in interpretation, especially when the responses are clarified by presentation as 'time slices'. Each slice is in effect a horizontal plot of response intensity over a depth slice calibrated from the estimated

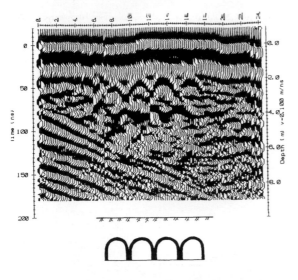

Fig. 127. This fine example of a pulseEKKO IV survey by the STG Corporation, in the Chang Jiang (Yangtse) valley upstream from Shanghai, revealed an ancient shelter made up of four contiguous air-filled vaulted chambers constructed from concrete. Their internal height and width were approximately 2 m, as was their depth below the surface. The ground consisted of hard silty clay underlain by sloping country rock. The frequency was 100MHz. The top scale is metres along the ground; the left-hand vertical scale is travel time for the reflected waves in nanoseconds (10^{-9} second); the right-hand scale is depth, assuming a velocity of 0.1 metre per nanosecond.

speed of transmission, and avoids the clutter of reflections from different depths and directions which can make the response pattern of a complete radar section notoriously complex and difficult to interpret. Good results have been obtained in Japan from barrows, and especially the tombs within them (Goodman and Nishimura, 1993), and from kilns (Goodman *et al.*, 1994, and references therein). Taking the latter as an example, a series of six 4-nanosecond time windows was used, equivalent to 8 cm (3 in) slices, down to a depth of 48 cm (18 in). Plotted as contours of reflection intensity, these clearly defined a kiln ranging in depth from about 28 cm (11 in) to at least 44 cm (17 in). Such data can also be plotted in three-dimensional perspective, as well as being corrected for variations in surface topography (Nishimura and Goodman, 1995). Good radar conditions, with soils giving consistent transmission velocities, are important in such work, and indeed in radar interpretation generally.

Seismic reflection and refraction

These standard long wavelength geophysical methods are unlikely to become major archaeological techniques, mainly because they are better adapted to larger-scale work; however, with continued refinement, and on suitable sites, they can (or will) produce useful results.

A careful optimisation study (Cross, 1994) has demonstrated that the seismic reflection technique has archaeological potential. In a practical experiment, the author placed two concrete blocks in a stepped trench at depths of 1 m (3.3 ft) and 2 m (6.6 ft), and backfilled the trench with homogeneous sand. Concrete blocks 2 m (6.6 ft) long and 0.5 m (1.6 ft) high were placed in line on the upper step (block 0.4 m (1.3 ft) wide) and the lower (block 0.6 m (2 ft) wide). Quite clear response to the upper block was obtained in a cross-traverse, which also indicated the limits of the trench, and in an axial traverse which clearly showed the change in level of the trench. Good images of the lower block do not seem to have been obtained, but it may be that the dimensions of the blocks were less suitable for detection than those of the containing trench.

Rather than the traditional sledgehammer, the best energy source was found to be .22 calibre blank cartridges manufactured for powder-actuated tools, and fired in a closed 'gun'. The detector was an OYO Eddy-Seis. For the concrete block experiments, the source-receiver separation was 30 cm (1 ft), ensuring that the direct signal interfered minimally with reflected signals from shallow features, and setting a minimum detection depth of 30 cm (1 ft). The sampling interval was 5 cm (2 in) – rather costly in cartridges and probably not necessary for larger features.

Preliminary field tests in Greece have produced some positive results, for instance the tracing of a harbour wall, and at the time of writing this work is scheduled to continue on a more substantial scale.

Seismic refraction, pioneered in early work on Etruscan tombs by the Lerici Foundation, has also been revived, notably because of its ability to give three-dimensional information (Ovenden-Wilson, 1994). Its most common function is in tracing the depth of bedrock beneath unconsolidated overburden. A pulse of energy, usually from a hammer blow, is delivered to the ground in line with a series of equally spaced geophones. Part of the seismic energy is refracted along the top of the bedrock, and some of this energy escapes through the soil to each geophone in turn. Once the seismic velocity in the overburden has been established, and the arrival intervals to each geophone recorded on a seismogram, the depth to bedrock is readily determined. Over level or gently shelving bedrock, travel times show a steady linear progression. If, however, the signal crosses a substantial archaeological feature cut into the rock, its arrival is delayed and the record displaced. Assuming that the overburden and filling are uniform, the depth of the feature can be estimated from the travel time.

Seismic refraction was used successfully to trace the central feature of the vallum, a substantial ditch parallel with Hadrian's Wall at Rudchester (*Vindobala*) (Goulty *et al.*, 1990; Goulty and Hudson, 1994). The ditch was suitably substantial: 8 m (26 ft) wide at the top, with a flat floor 2.5 m (8.2 ft) wide and a depth of 3 m (9.8 ft), cut into sandstones and shales of the Upper Carboniferous Millstone Grit series. The overburden was glacial drift.

Chapter 7 – Interpretation and presentation

Relief plots

A valuable addition to modern data-processing packages is the facility to generate relief plots. These are envisaged as being produced by an imaginary artificial sun which can be manipulated to shine on the data from a variety of directions and elevations, with Geoplot 2.01 for instance at 45° intervals, and a standard elevation of 35°. A halftone or dot-density display is used. Any peaks or depressions, or lineations with a component perpendicular to the sun direction, are apparently illuminated on the side facing the sun, and shadowed on the other side. Linear features aligned with the sun are invisible, and the sun must be stepped around the dataset to find the direction in which it will illuminate all the significant features; or if this is not possible more than one plot may need to be produced.

Relief plots often prove to be the most sensitive and effective way of presenting a survey. They also provide a means of making an initial examination of survey data without requiring any decisions about plotting levels or filtering. A good example of this is shown in Fig. 128: plots of a small twin-electrode

resistivity survey of two partial 30 m (98.4 ft) squares, obstructed by an irregular line of vegetation on the right-hand side. The geology is sand underlain by clay. Readings are at 1 m (3.3 ft) interval, with interpolation.

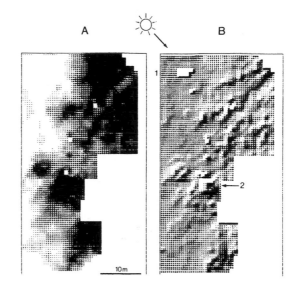

Fig. 128. Effect of artificial sun relief plotting.

A is a straightforward initial grey-scale plot over the range -1 to +1 standard deviations = white to black, produced on a dot matrix printer. A stream runs just outside the edge of the survey to the left, with associated damp ground of low resistivity contrasting with the dry, higher ground to the right, so that detail is lost on both sides. The only artificial-looking feature is the weak low-resistance diagonal line running toward the top right. In the relief plot, B, the background slope is automatically rectified and the picture comes to life. Linear features, probably trackways, fan out from the top right, and a small oblong feature, probably a building, is highlighted at 2. The blank area 1, top left, and other small blanks are where the survey was interrupted by trees and other obstacles.

Automatic interpretation

Now that large datasets can be produced and displayed with ease, it becomes commensurately difficult to interpret some of the data, partly because the plots have to be printed on quite small scales in order to be manageable. Many ditched features are clear enough, but the appearance of other anomalies seems to fall between a scatter of individual features and simply rubbish-induced noise. Objective criteria are needed to separate these, and to decide on the credibility level of apparent sites. Possible approaches range from traditional pattern recognition, as used in satellite surveillance (Roberts, 1995), to the use of artificial neural networks, which would enable a system to 'learn' and recognise significant anomaly shapes. It has already been noted above that dipole pattern matching software, which will produce maximum likelihood estimates of object size, depth and possible identification is being developed for the Geometrics G-858 caesium magnetometer.

Additional references for the second edition

ALLEN, M.J. and M. J. MACPHAIL, R. I., 1987. 'Micromorphology and magnetic susceptibility studies: their combined role in interpreting archaeological soils and sediments.' In Fedoroff, N., Bresson, L. M. and Courty, M.-A. (eds) *Micromorphologie des Sols*, 669–676. Association du Sol Français.

BARKER, P. and CROWTHER, J., 1995. 'Magnetic susceptibility: sorting out the archaeology from the geology.' Archaeological Prospection Conference, Bradford.

BEVAN, B. W., 1991A. Technical report: 'selecting a magnetometer'. *Society of Archaeological Sciences Bulletin* 14(4), 2–5.

BEVAN, B. W., 1991b. 'The search for graves.' *Geophysics* 56, 1310–1319.

BEVAN, B. W., 1993. Technical report: 'selecting a resistivity meter'. *Society of Archaeological Sciences Bulletin* 16(3), 2–7.

BEVAN, B. W. and KENYON, J., 1975. 'Ground penetrating radar for historical archaeology.' *MASCA Newsletter* 11(2), 2–7.

CARR, C., 1982. *Handbook on Soil Resistivity Surveying*. Center for American Archeology Press, Evanston.

CLARK, A. J., 1992. 'Archaeogeophysical prospecting on alluvium.' In Needham, S. and Macklin, M. G. (eds) *Alluvial Archaeology in Britain*, 43–49. Oxbow, Oxford.

CROSS, G. M., 1994. 'Re-evaluation of the seismic reflection method for archaeological application.' Ph.D. thesis, University of British Columbia, unpublished.

DAVID, A., 1995. *Geophysical Survey in Archaeological Field Evaluation.* English Heritage, London.

DEUEL, L., 1969. *Flights into Yesterday*. Macdonald, London.

FASSBINDER, J. W. E., 1994. *Die magnetischen Eigenschaften und die Genese ferrimagnetischer Minerale in Böden*. Verlag Marie L. Leidorf, München.

FASSBINDER, J. W. E., STANJEK, H. and VALI, H., 1990. 'Occurrence of magnetic bacteria in soil.' *Nature* 343, 161–163.

FRENCH, C. A. I., 1994. *Excavation of the Deeping St Nicholas Barrow Complex, South Lincolnshire*. Heritage Trust of Lincolnshire, Sleaford.

GAFFNEY, C. and GATER, J., 1993. 'Practice and method in the application of geophysical techniques in archaeology.' In Hunter, J. and Ralston, I. (eds) *Archaeological Resource Management in the UK*, 205–214. Alan Sutton/Institute of Field Archaeologists, Stroud.

GOODMAN, D. and NISHIMURA, Y., 1993. 'A ground radar view of Japanese burial mounds.' *Antiquity* 67, 349–354.

GOODMAN, D., NISHIMURA, Y., UNO, T. and YAMAMOTO, T., 1994. 'A ground radar survey of medieval kiln sites in in Suzu city, western Japan'. *Archaeometry* 36, 317–326.

GOULTY, N. R., GIBSON, J. P. C., MOORE, J. G. and WELFARE, H., 1990. 'Delineation of the vallum at Vindobala, Hadrian's Wall, by a shear wave seismic refraction survey.' *Archaeometry* 32, 71–82.

GOULTY, N. R. and HUDSON, A. L., 1994. 'Completion of the seismic refraction survey to locate the vallum at Vindobala, Hadrian's Wall.' *Archaeometry* 36, 327–335.

GRARD, R. and TABBAGH, A., 1991. 'A mobile four electrode array and its application to the electrical survey of planetary grounds at shallow depths.' *Journal of Geophysical Research* 96(B3), 4117–4123.

GRIFFITHS, D. H. and BARKER, R. D., 1994. 'Electrical imaging in archaeology.' *Journal of Archaeological Science* 21, 153–158.

HERBICH, T., 1993A. 'The method of estimation of the extent of the mining field of flint mines through observation of the arrangement of surface layers.' *Archeologia Polski* 38, 23–35.

HERBICH, T., 1993B. 'The variations of shaft fills as the basis of the estimation of flint mine extent: a Wierzbica case study.' *Archaeologia Polona* 31, 71–82.

HERBICH, T., MISIEWICZ, K. and SOMMER, C. S., 1993. 'Geophysical prospection in Roman Rottweil – *Arae Flaviae*.' *Fundberichte aus Baden-Württemberg* 18, 83–111.

MAEKAWA, K., SAKEI, H., UNO, T. and KANER, S., 1995. *Swavesey: Geophysical Survey at Blackhorse Lane, 1994: Interim Report.* Toyama and Cambridge Universities.

MCGARTH, S. P. and LOVELAND, P. J., 1992. *The Soil Geochemical Atlas of England and Wales*. Blackie Academic and Professional, Glasgow.

NISHIMURA, Y. and GOODMAN, D., 1995. 'Static corrections of GPR time slices in archaeological prospection.' Archaeological Prospection conference, Bradford.

NOËL, M. and XU, B., 1991. 'Archaeological investigation

by electrical resistivity tomography: a preliminary study.' *Geophysical Journal International* 107, 95–102.

OVENDEN-WILSON, S. M., 1994. 'Application of seismic refraction to archaeological prospecting.' *Archaeological Prospection* 1(1), 53–63.

ROBERTS, K., 1995. 'The archaeological applications of geophysical survey techniques.' Ph.D. thesis, University of Cambridge, unpublished.

SCOLLAR, I., TABBAGH, A., HESSE, A. and HERZOG, I., 1990. *Archaeological Prospecting and Remote Sensing.* Cambridge University Press, Cambridge.

TANAKA, Y., 1995. 'An archaeogeophysical survey of San Xia, People's Republic of China.' Archaeological Prospection Conference, Bradford.

TABBAGH, A., 1984. 'On the comparison between the magnetic and electromagnetic prospection methods for magnetic features detection.' *Archaeometry* 26(2), 171–182.

TABBAGH, A., 1993. 'Determination of the electrical properties of the ground at shallow depth with an electrostatic quadrupole.' *Geophysical Prospecting* 41, 579–597.

General reference

DEARING, J., 1994. *Environmental Magnetic Susceptibility: Using the Bartington MS2 System.* Bartington Instruments, Witney

GAFFNEY, C., GATER, J. and OVENDEN, S., 1991. *The Use of Geophysical Techniques in Archaeological Evaluations.* Technical Paper No. 9. Institute of Field Archaeologists, Birmingham.

GURNEY, D. A., 1985. *Phosphate Analysis of Soils: a Guide for the Field Archaeologist.* Technical Paper No. 3. Institute of Field Archaeologists, Birmingham.

HEIMMER, P. G. and DE VORE, S. L., 1995. *Near-Surface, High Resolution Geophysical Methods for Cultural Resource Management and Archeological Investigations* (Revised edition). Interagency Archeological Services, Denver.

TELFORD, W. M., GELDART, L., SHERIFF, R. E. and KEYS, D. A., 1977. *Applied Geophysics.* Cambridge University Press, Cambridge.

Archaeological Prospection. This journal is a recently established successor to Fondazione Lerici's *Prospezioni Archeologiche.* It is published by Wiley at Chichester and edited in the Department of Archaeological Sciences, University of Bradford.

The references quoted above as deriving from the 1995 Archaeological Prospection Conference will probably appear in *Archaeological Prospection.*

Supplementary index

Bacterial magnetism, 182
Carbolite furnace, 180
Chang Jiang (Yangtse) River, China, radar and other surveys, 183
Conductivity detectors
 EM31, 183
 EM38, 171, 183
Coventry University, 182
Deeping St Nicholas, Cambridgeshire, susceptibility survey, 182
Evaluation surveys
 use of magnetic susceptibility in, 175ff
 types, 177f
Electrical imaging, 171
Fermentation, 101, 109, 182
Flint mine surveys, 171
Geoplot 2.01 software, 178, 185
Hadrian's Wall, seismic refraction survey of vallum, 185
Holland, magnetic susceptibility of soils, 175
Hurst Park, Surrey, magnetic susceptibility survey, 178
Interpretation
 automatic, 185f
 relief plotting, 185
Magnetometers
 Caesium G858, 175
 Overhauser GSM19, 174f
Northamptonshire survey, 177
Northborough by-pass, Cambridgeshire, magnetic susceptibility survey, 182
Overton Down experimental earthwork, Wiltshire, magnetic susceptibility study, 176
OYO Eddy-Seis geophone, 184
Poland, flint mine surveys, 171
Pulse-EKKO ground radar equipment, 183
Reigate, Surrey, survey at Priory, 172f
Resistivity
 electrostatic contacts, 173
 pole-pole configuration, 173
 site analysis and garden surveys, 172f
 theory, 173f
Seismic
 reflection, 184f
 refraction, 185
Soil Survey and Land Research Centre, 182
Susceptibility, magnetic
 Bartington sensors, 179
 Eijkelkamp sampler, 179
 evaluation surveys, 175ff
 heating tests for human effects, 179ff
 in excavation, 182
 Overton Down study, 176
 soil map of England, 182
 WSL-B meter, 182

Index

Acoustic reflection, 121
Aitken, Martin J., 16ff, 71
Al Chalabi, M. M., 48
Alldred, John, 19, 21ff
Allen, Michael, 115
Allen, G. W. G., 11
Alluvium, 114, 131; *see also* Silt
Ancient Monuments Laboratory (AML), 7, 20, 72, 78, 92, 110, 130, 132f, 158
Andover Archaeological Society, 139
Antiferromagnetism, 100
Apple computer, 25, 149ff
Ashmolean Museum, Oxford, 12
Aspinall, Arnold, 22
Atkinson, Richard J. C., 11, 59, 142ff, 162
Automatic recording systems, *see* Fluxgate gradiometers, Proton magnetometers *and* Resistivity meters
Avebury, Wiltshire, 45, 72, 130

Bacteria, 101
Badbury Rings, Dorset, 80
Balaam, Nick, 112
Banjo,
    enclosure, 88, 109ff, 139
    instrument, *see* SCM
Bartington MS2 magnetic susceptibility meter, 102
    Sensors: MS2B, 102, 115; MS2C, 102; MS2D, 104, 111, 116; MS2F, 104ff, 114ff; MS2H, 105
Bartlett, Alister D. H., 7, 146, 155ff, 78
Barton Stacey, Hampshire, magnetic survey of, 133, 136ff
Basalt, 65; *see also* Igneous rocks
Belshé, John, 16
Birmingham University, 61
Bodmin Moor, 92
Boscombe, Wiltshire, 152ff
Bosing, 11, 121
Bowen, H. C., 110
Bradford University School of Archaeological Sciences, 22, 59
Braughing, Hertfordshire, 110
Briar Hill, Northampton, 72
Bristol, Druid Stoke, 142ff
British Gas PLC, 133
British Museum Research Laboratory, 120

Brodgar, Ring of, Orkney, 91
Bronze Age sites, 112ff, 128, 135, 142, 156; *see also* Hog's Back bell barrow
Building foundations, 124ff; *see also* Walls
Burton Fleming, North Humberside (Yorkshire Wolds), 20ff, 144, 150ff
Butser Ancient Farm Research Project, 101, 109, 112

Cadbury Castle, Somerset, 22
Calcite, 55
Calcium detection, 121
Canada, 121
Carboniferous, 92; *see also* Limestone
Caribbean, *see* St Lucia
Centre de Recherches Géophysiques, Garchy, 25, 60
Chalk, 109, 142, 153
    magnetic properties, 80, 87ff, 92, 100
    resistivity properties, 49ff, 58
Chemical prospecting, *see* Geochemical
Chilterns, 92
Clark, Perry D., 7
Clay, 44
    Boulder, 55, 94
    London, 53ff, 139
    resistivity, 27,
    thermoremanence, 64ff
    -with-flints, 58, 87ff, 92, 100
Cleveland, 92
Climate, effect on resistivity surveying, 48ff
Clyde, River, 97
Colani, C., 22
Colour plotting, 147, 150
Combretovium Roman town, Baylham House, Suffolk, 89
Compounds, magnetic, in soils, 100
Compton, Surrey, 51, 128
Conductivity detection, 34ff
Coneybury henge, Wiltshire, magnetic susceptibility survey of, 110ff, 156
Contour plots, 141ff
Cotswold Hills, 92, 116, 129, 135
Cottam, North Humberside, 55, 144, 150ff
Cretaceous, 92
Crete, 92
Crop marks and resistivity, 12, 15, 52f
Cumbria, Little Hawes Water, 115

Cunetio Roman town (Mildenhall), Wiltshire, 14, 59
Curie point, 65
Dainton, Devon, magnetic susceptibility survey at, 112ff, 156
Dartmoor, 92
Darwin, Charles, 114
David, Andrew E. U., 7
Devonian, 92
Distillers Company, 14
Ditches, 124ff
    magnetic susceptibility response, 125ff
    magnetometer response, 78ff, 97, 125ff
    phosphate response, 126ff
    resistivity response, 13, 15, 48ff, 58, 124ff 150; ring-, 59
Diurnal variation, magnetic, 66ff
Dolerite, 114
Dorchester-on-Thames, 11, 12
Dorset, 92
Dot density, 143ff
Dowsing, 123
Drayton, Oxfordshire, 115
Druid Stoke, *see* Bristol
Durobrivae Roman town (Water Newton), Cambridgeshire, 16
Durrington Walls, resistivity tests, 49, 51, 54

Earth's magnetic field, *see* Geomagnetic field
East Anglia, 92
Egypt, 120ff
Electrolytic tank, 37ff, 57
Elsec (Littlemore Scientific Engineering Company), 71
Elslack, Cumbria, 35
EMI, 21ff
English Heritage (Historic Buildings and Monuments Commission for England), 112, 149
Epson HX20 portable computer, 25, 77ff
Equator, 64, 83
Etruscan tombs, 17, 121
Evans, J. G., 49

Farnham, Surrey, 150
Fasham, Peter, 110
Fermentation, 101, 109
Ferrimagnetism, 100
Ferromagnetism, 100
Field, Norman, 139
Field walking, use of magnetic scanning, 89
Filtering, 150ff
    band-pass, 154ff
Fluxgate gradiometers, *see also* Magnetometers
    anomaly shapes and latitude effects, 82f
    computerized recording systems, 73, 77ff
    continuous recording system, 71ff
    depth estimation with, 90
    depth sensitivity, 78ff

early use, 19, 23ff
    Geoscan instruments, 70ff, 77ff, 81, 88
    Philpot instruments, 70ff
    principles and design, 69ff
    scanning with, 83, 85ff
    spatial resolution, 80ff
    surveys with, 132ff
Foraminifera, 55
Foster, Eric J., 21ff
France, 54
Freezing, effect on soil resistivity, 54
Friesinger, Herwig, 116

Geochemical prospecting methods, 120ff
Geologger, 61
Geology, *see* Alluvium, Chalk, Clay, Gravel, Igneous rocks, Jurassic, Limestone, Sand and sandstone, Millstone Grit
    magnetic surveys, effect on, 92ff
    resistivity surveys, effect on, 48ff
Geomagnetic field, 64, 67, 82ff
Geomorphology and archaeological potential, 130ff
Geonics electromagnetic instruments
    EM31, 34ff, EM38, 34ff, 61, 105
Geoscan Research, *see* Fluxgate gradiometers *and* Resistivity meters
Geospace Consultancy Services Ltd., 119
Glasgow, 97
Gleying, 100, 114,
Goethite, 100
Goodhugh, P., 53
Gorman, Michael, 120
Graham, Andrew *and* David, 150
Granite, 92
Grassland, magnetic susceptibility surveys on, *see* Susceptibility, magnetic
Gravel, 54, 92, 124ff, 139, 156
Graves
    EM31 response, 36
    Etruscan, *see* Etruscan tombs
    magnetometer response, 98
    resistivity response, 151
Greece, 92, 121
Grids, *see* Survey grids
Grime Graves, Norfolk, 58, 154ff
Gurdler, John, 22

Habberjam, G., 46
Haddon-Reece, David, 7, 24
Haematite, 64ff, 100ff
Hall, Edward T., 16, 22
Hampshire, 92; *see also* Barton Stacey, M3, Winchester
Handley Down, Dorset, 11
Henge monuments, 11; *see also* Avebury, Brodgar, Coneybury, Durrington Walls, Stenness *and* Stonehenge

Hesse, Albert, 27, 53ff
Highland Zone, 100
    magnetic surveying in, 92ff
Hillforts, 59, 86ff
    Anstiebury, Hascombe *and* Holmbury, Surrey, 87
    Maiden Castle, Dorset, 146ff
Hog's Back, Surrey, bell barrow, 49ff, 54ff, 57
Howell, Mark, 105ff

Ice, *see* Freezing
Igneous rocks, 92ff, 116ff, 133; *see also* Basalt *and*
    Susceptibility, magnetic
Illinois, 115, 121
Insall, G. S. M., 53
Iona, 96ff
Iron Age sites, 20ff, 55, 59, 80, 88, 110, 133, 144,
    136ff, 138ff, 147ff, 150f; *see also* Hillforts
Isograph plots, *see* Contour plots
Italy, 92; *see also* Lerici

Japan, 61, 63, 119, 121
Jordan, Bab ed-Dhra, 36
Jurassic, 92, 135, 139

Keuper Marl, 92
Kilns
    magnetic susceptibility response, 125ff
    magnetometer response, 65, 74ff, 80f, 125ff
    phosphate response, 126ff
    resistivity response, 125ff

Laming, Annette, 12
Laser printer, 147
Latitude, effect on magnetic anomalies, *see*
    Magnetometers *and* Fluxgate gradiometer
Latton, Wiltshire, 145ff
Leapfrog operating method, *see* Resistivity electrode
    (probe) configurations
Leatherhead, Surrey, 20
Le Borgne, E., 17, 100ff, 106
Lepidocrocite, 100
Lerici, Fondazione, 17, 121
Limestone, resistivity properties, 53ff
    Carboniferous, *see* Bristol
Linington, Richard E., 17
Linwood, Hampshire, 23
Littlemore Scientific Engineering Company, *see* Elsec
Littler, N. S., 107
Liverpool University, 115
Llawhaden, Dyfed, 76

M3, Hampshire, *see* Motorway routes
Maghaemite, 64
Magnetite, 64ff, 103
Magnetometers
    alkali vapour or optical pumping, 19, 68ff

anomaly shapes and latitude effects, 82ff
automatic recording, 19; *see also* Fluxgate
    gradiometer
depth sensitivity, 78ff
differential, 19, 67
gradiometers, *see* Fluxgate gradiometer *and* Proton
    gradiometer
metal detection with, 98
proton, 16ff, 66ff
response to archaeological features, 90, 92, 125ff
spatial resolution, 80ff
Magnetostratigraphy, 115f
Maiden Castle, Dorset, 146ff
Martin, John, 14
Mason, R. W. I., 25
Medieval and later sites, 150, 156
Megger Earth Tester, *see* resistivity meters
Metal detectors, 121f
Meteorological Office, 55
Mexico, Gulf of, 120
Microfilm plotter, 146
Middle East, 121; *see also* Egypt, Jordan, Oman
Millstone Grit, 122
Ministry of (Public Building and) Works, 60, 71
Mississippi, River, 121
Motorway routes, scanning, 87ff
M-Scope, Fisher, 106
Mullins, Christopher, E., 21ff, 101
Munday, Charles W., 14
Musty, John W. G., 20

National Grid, 158ff
Natural Environment Research Council (NERC),
    131
Neatham, Hampshire, 74, 97
Nene, River, 92
Neolithic sites, 94ff, 72, 115, 129, 142ff, 154f; *see also*
    Henge monuments
Netherlands, 131
Nigeria, 83

Oceanfix International Ltd., 120
Oman, 34, 36, 72, 83ff
Ordnance Survey, 158
Ordovician, 92
Orkney, 93ff, 133
OYO Corporation, 61, 63, 119

Painshill Park, Surrey, 55
Palmer, L. S., 47ff
Paramagnetism, 100
Peat, 131
Pedogenic processes, 101
Pennsylvania, 131;
    University of, Applied Science Center, 19, 23
Pewsey, Wiltshire, 97ff

pH, detection by, 121
Philpot, Frank V., 23
Philpot Electronics Ltd., *see* Fluxgate gradiometers
Phosphate detection, 108ff, 112ff, 120ff
  early use, 11
  response to archaeological features, 126ff
Pits *and* quarries, 59, 78ff, 90, 124ff, 142
  magnetic susceptibility response, 125ff
  magnetometer response, 78ff, 90, 125ff
  phosphate response, 126ff
  resistivity response, 124ff
Pitt Rivers, Lieut. General A., 11
Plessey Company, 23
Plotting methods
  colour, 147ff
  grey-scale, 147
  manual, 133
  pseudo-relief, 147, 150
  symbols, 156ff
  traces, 132ff
Polarization
  induced, as a detection method, 122
  problems in resistivity measurement, 29ff
Pouilly, Nièvre, 54
Profiling, resistivity, 61
Proton gradiometer, 16f, 68;
  Maxbleep, 19
Psion Organiser, 102
Pulsed induction meter (PIM), 22, 106, 109
Pyramids, *see* Egypt

Quarries, *see* Pits

Radar
  ground-penetrating, 26, 118ff
  Space Shuttle, 120
Radley, Oxfordshire, 135
Ralph, Beth, 19
Rees, A. I., 48
Research Laboratory for Archaeology, Oxford, 16f,
  19, 22
Resistivity, DC, 36, 63
Resistivity electrode (probe) configurations
  continuous contact, 59ff
  double dipole, 38ff, 44
  dual spacing, 155ff
  leapfrog operating method, 13, 30ff
  multi-probe, 60ff
  Palmer, 38ff, 47ff
  Schlumberger, 38ff, 47
  spacing, optimization, 57f; effect of errors, 58
  square, 38ff, 46f, 128, 142, 144
  twin, 38, 41, 44, 47ff, 143
  Wenner, 37ff,
Resistivity meters
  automatic, 33ff
  automatic recording with, 44f, 59ff

constant current, 33f
early use, 12ff
Geoscan RM4, 33f, 36, 44ff
manual balance, 30ff
Martin-Clark, 19, 30f, 34, 44
Megger Earth Tester, 12, 33
Resistivity, new developments
  profiling, 61ff
  rapid survey methods, 59ff
  tomography, 61ff
Resistivity response to archaeological features, 124ff
Rheinisches Landesmuseum, 19
Rhineland, 18ff
Richards, Julian, 110
Roman and Romano-British sites, 14, 16, 23, 35, 44,
  48, 51, 53ff, 55, 57ff, 59, 62ff, 74, 89, 97, 110,
  120, 128, 134, 139ff, 145ff, 152ff
Rothamstead, Hertfordshire, 54
Royal Commission on the Historical Monuments of
  England, 107, 149

St Lucia, Windward Islands, 94, 116ff
Salisbury Plain, 92
Sand *and* sandstone, 139
  magnetic properties: Lower Greensand, 86ff, 92;
    Old Red, 94;
  magnetic scanning, 86ff;
  resistivity properties, 124, 128, 130; Triassic, 48,
    54ff
Saxon sites, 22, 88ff, 97ff, 110, 120, 128, 130
Scanning, magnetic, *see* Fluxgate gradiometer *and*
  Sandstone
Schwarz, G. T., 22
SCM (Soil Conductivity Meter or "Banjo"), 22, 105ff
Scollar, Irwin, 19, 143, 146
Seismic reflection profiling, 120
Selwood, Brian, 114
Servomex field plotter, 29
Servoscribe M plotter, 24, 71
SH3 electromagnetic instrument, 105
Sharpe Howes, Yorkshire Wolds, 142
Sheffield University, 61
Silt, 92
Skara Brae, Orkney, 94ff
Smith, George, 112
Smoothing, 153
Soil marks, 110
Soils
  magnetization, 65ff, 92, 96, 100ff
  resistivity, 27
Somers, Lewis E., 77f
South Downs, Sussex, 115
Sowerbutts, W. T. C., 25, 77
Spain, 38
Stead, Ian M., 20
Stenness, Standing Stones of, Orkney, 93f, 112

Stone, *see also* Igneous rocks *and* Walls
  resistivity 27,
  thermoremanence, 65
Stone, Kent, 89
Stonehenge, 100, 110, 112
Strata Software, 150
Sudan, 120
Superparamagnetism, 103
Surrey University, resistivity tests, 54
Survey grids, 71, 158ff
Susceptibility, magnetic, 65ff, 99ff
  fermentation effect, 101
  frequency dependent, 102ff
  grassland surveys, 114ff
  heating effect, 101
  igneous area surveys, 116ff
  instruments, 101ff
  magnetostratigraphy, 115f
  response to archaeological features, 125ff
  topsoil survey, detailed, 110
Sutton Hoo, Suffolk, 110, 120
Swan, Vivien, 25
Swindon, Wiltshire, Groundwell Farm, 138f
Sybaris, Italy, magnetometer survey, 19

Tabbagh, Alain, 27, 105
Tadworth, Surrey, magnetic susceptibility and
    phosphate experiment, 107ff, 112, 156
Temperature, effect on soil resistivity, 54
Terra rossa, 92
Tertiary, 92
Thame, Oxfordshire, resistivity survey, 156
Thames, River, 92, 116, 135
Thermal sensing, 122
Thermoremanent magnetization, 64ff
Tite, Michael S., 21ff, 92
Tomography, 61ff
Topsoil survey, *see* Susceptibility, magnetic,
    Phosphate *and* Geochemical prospecting
    methods
Toshiba computer, 73, 78
Tropics, 92
Trust for Wessex Archaeology, 110
Turkey, 92

Turton, Basil, 146

United States of America, 67, 115, 120
Units
  magnetic, 64
  magnetic susceptibility, 99f
  resistivity, 27
Verulamium, St Albans, Hertfordshire, 62ff
Vienna State University, 116
VLF (very low frequency radio transmissions), 36

Wainwright, G. J., 49
Wales, 92; *see also* Llawhaden
Walker, Roger, 22
Wall, Staffordshire, resistivity tests, 48, 54
Walls, *see also* Stone
  magnetometer response, 147, 125ff
  resistivity response, 14ff, 44, 55, 57f, 124ff, 151ff
  radar response, 120
Wansdyke, 58
Water Newton, *see Durobrivae*
Webster, Graham, 16, 139
Weekley, Northants, 80
Welland, River, 92
Wenner, Frank, 37
Wessex, 110
Weymouth, John W., 26, 121
Wharram Le Street, Yorkshire, 134
Wheeler, R. E. M., 149
Wimborne Minster, Dorset, Lake Farm, 139ff, 55
Winchester, 58f, 110f
Windward Islands, *see* St Lucia
Woodhenge, Wiltshire, 53
  resistivity and conductivity tests, 49ff
Worms, 114
Wraysbury, Berkshire, 128, 130

Xanten, Rhineland, 18ff

York, 119ff
Yorkshire Wolds, *see also* Burton Fleming, Cottam,
    Sharpe Howes
  magnetic response, 92
  resistivity response, 55, 142